MICROMACHINED THIN-FILM SENSORS
FOR SOI-CMOS CO-INTEGRATION

Micromachined Thin-Film Sensors for SOI-CMOS Co-Integration

by

Dr. J. LACONTE
*Intersema Sensoric SA,
Bevaix, Switzerland*

Prof. D. FLANDRE
*Université Catholique de Louvain,
Louvain-La-Neuve, Belgium*

and

Prof. J.-P. RASKIN
*Université Catholique de Louvain,
Louvain-La-Neuve, Belgium*

A C.I.P. Catalogue record for this book is available from the Library of Congress.

ISBN-10 0-387-28842-2 (HB)
ISBN-13 978-0-387-28842-0 (HB)
ISBN-10 0-387-28843-0 (e-book)
ISBN-13 978-0-387-28843-7 (e-book)

Published by Springer,
P.O. Box 17, 3300 AA Dordrecht, The Netherlands.

www.springer.com

Printed on acid-free paper

All Rights Reserved
© 2006 Springer
No part of this work may be reproduced, stored in a retrieval system, or transmitted
in any form or by any means, electronic, mechanical, photocopying, microfilming, recording
or otherwise, without written permission from the Publisher, with the exception
of any material supplied specifically for the purpose of being entered
and executed on a computer system, for exclusive use by the purchaser of the work.

Printed in the Netherlands.

Abstract

While in some applications such as gas-flow sensors, the co-integration of a sensor with its surrounding electronics on a single chip is just an asset, co-integration is inescapable in other cases such as transistors-based pressure sensors or low-loss microwave circuits insulated from the substrate, on a membrane. Moreover, co-integration performed in the most advanced of the Complementary-Metal-Oxide-Semiconductor (CMOS) technologies to date -namely Silicon-on-Insulator (SOI) technology- provides many significant benefits regarding the performance, reliability, miniaturization and processing easiness without significantly increasing the final cost.

Thin dielectric membranes constitute the starting material for a large number of sensors thanks to their ability to act as a mechanical support or an electrical and thermal insulator. We particularly focused on the thermal insulation feature to build fully CMOS-SOI co-integrated gas-flow sensors. A one-μm-thick robust and flat dielectric multilayered membrane has been built taking precisely into account the residual stresses in each constitutive layer. A complete review and summary of the main concepts of thin film mechanics is detailed for an in-depth understanding. A new measurement methodology based on substrate curvature and deflected microstructures has been developed in order to accurately quantify the residual stresses in each layer and in their stacking. To release the membrane in post-processing, the Tetramethyl Ammonium Hydroxide (TMAH) silicon micromachining solution has been used and optimized in order to increase its selectivity towards aluminum. This particular wet etching technique was extensively reviewed and enhanced by our own experiments, in particular for assuring CMOS compatibility.

A novel loop-shape polysilicon microheater implementing the basic heating cell of new simple gas-flow sensors has been designed and produced in a CMOS-SOI standard process. High thermal uniformity, low power consumption and high working temperature have been targeted and confirmed by extensive measurements. In particular, the electrical properties of polysilicon versus temperature and annealing time have been analyzed in depth. The gas-flow sensor has been optimized to be integrated in intermediate- and post-processing of a standard CMOS-SOI fabrication. The thermopiles of the flow sensor as well as the interdigitated electrodes of the gas sensor have judiciously been chosen in this purpose. The sensing films of the gas sensors consisted in sputtered and drop coated metal oxide layers such as SnO_2 and WO_3. Measurements in the presence of a nitrogen flow and gas revealed a fair sensitivity on a large flow velocity range for the flow transducer, as well as a good sensitivity to gases such as ethanol, ammonia and nitrogen dioxide for the gas transducer. The whole process has confirmed its full CMOS compatibility by measuring MOS transistors, capacitors and gated diodes on the same chip as the sensors after each post-processing step.

Finally, transistors integrated in small silicon islands located in the middle of our dielectric membrane have been studied and presented as a concluding demonstrator of the co-integration in SOI technology. Such devices open the door to numerous new applications where integrated circuits and sensors are merged in order to target higher performance in harsh environments.

Acknowledgments

The authors are grateful to the external members of Mr. Laconte's PhD jury, Professors Dominique Collard, Eduard Llobet and Nico de Rooij who gave judicious advices to put the final touch to the researches and the manuscript.

Large parts of the work presented in this book are the result of collaborations with the technical staff of UCL clean-rooms. We would really like to thank André Crahay, Christian Renaux, David Spôte, Nathanaël Mahieux for their devotedly help for processing the unrewarding steps of the fabrications, without forgetting Bohdan Katschmarskyj for its occasional implantations. Pierrot Loumaye and Miloud Zitout were also of great help to perform the incalculable wafer dicing and sensors packaging. We would also like to thank Laurent Morelle and Miloud Zitout for their electrical realizations as well as Pierrot Loumaye for his numerous mechanical manufacturings.

The students supported during their master theses were also essential to the accomplishment of this work. We would like to thank Cédric Dupont, Valérie Wilmart and Bertrand Rue. The help of Sébastien Jorez from the optical Laboratory of the university was also invaluable to me. He devoted a large part of its post-doctoral research to perform all the needed optical measurement set-ups, with a great patience and always, with a lot of efficiency. Thomas Pardoen, Yannick Bertholet and Joris Proost from the IMAP laboratory of the university were also of great help for their numerous advices in the mechanics of materials. Peter Tsolov Ivanov, thank you for your gas sensors measurements in Tarragona. We would also thank other UCL researchers, especially Bertrand Parvais, François Iker, Rémi Charavel, Luis Moreno, Rémi Pampin, David Levacq, André Tuor, Loïk Gence, Aryan Afzalian, Valeria Kilchytska and Alexandru Vlad who collaborated in the fabrications, designs or even in the reading of some parts of this manuscript. Many thanks for your pleasant company.

We also thank the computing staff of our laboratory which solved the incredible number of little problems I faced on Windows or Unix environments, Brigitte Dupont, Geoffroy

Simon, Damien Giry and Laurent Vancaillie. Moreover, we cannot thank enough our secretaries, Anne Adant and Annette Kreuwels and our accountant, Marie-Hélène Dewael for the numerous urgent orders, always placed at the last minute.

Our final thoughts go to Axelle for her help to correct the introduction and conclusions of this work. Jean also especially thanks his wife, Bénédicte, who immensely supported him in his daily investments for his thesis and who encouraged him to finalize it despite of the organization of their wedding at the same time...

Denis and Jean-Pierre have also grateful thoughts for their wives and children who sometimes endure their long working hours and regular absences.

Contents

I Introduction: Context and motivations — 1
    0.1 Why Silicon-on-Insulator technology ? 4
    0.2 Why a thin-film membrane ? 6
    0.3 Why co-integration and CMOS compatibility ? 8
    0.4 Contents of the work 10

II Techniques and materials — 15

1 Silicon bulk micromachining with TMAH — 17
    1.1 Introduction 17
    1.2 Generalities about silicon micromachining 19
    1.3 TMAH silicon etching 23
    1.4 Selectivity versus dielectrics 28
    1.5 Selectivity versus aluminum 29
        1.5.1 State of the art 29
        1.5.2 Experimental results 32
        1.5.3 Summary of the TMAH etching steps in presence of Al 36
    1.6 Selectivity versus other metals 37
    1.7 Etch-Stop 38
    1.8 Undercutting 41
        1.8.1 Generalities 41
        1.8.2 Membranes patterning 43
    1.9 Summary 45

2 Thin dielectric films stress extraction — 47
    2.1 Introduction - Definitions 48
        2.1.1 Stress, strain and Elastic constants 48
        2.1.2 Uniform and non-uniform stresses and strains in thin films 52

		Extrinsic stress	53

Extrinsic stress 53
Intrinsic stress 55
2.2 Stress measurements by substrate
curvature method 56
    2.2.1 Theory 56
    2.2.2 Experimental results 61
    2.2.3 Discussions and summary 68
2.3 Strain measurements using micromachined
structures 69
    2.3.1 Clamped-clamped beam and ring-and-beam
          structures analysis 71
          Theory 71
          Experimental setups and results 76
    2.3.2 Microgauges analysis 82
    2.3.3 Cantilever beams analysis 86
          Monolayer cantilever beams 86
          Multilayer cantilever beams 94
          Discussion on multilayer membranes 96
    2.3.4 Summary and outlook 97
2.4 Final conclusions 102

III Microsensors   105

1 Low power microhotplate as basic cell   107

1.1 Introduction 107
1.2 Motivations 110
    1.2.1 Gas sensors 110
    1.2.2 Other sensors 111
1.3 Materials selection 111
    1.3.1 Membrane 111
    1.3.2 Microheater 112
1.4 Thermal design 113
    1.4.1 Basis of heat transfer theory 113
          Conduction 113
          Convection 114
          Radiation 115

		1.4.2	Thermal simulations	115

- 1.5 Device fabrication 120
    - 1.5.1 On bulk silicon substrate 120
    - 1.5.2 On SOI substrate 122
    - 1.5.3 Dicing and packaging 126
- 1.6 Microheater characterization and results 127
    - 1.6.1 Introduction 127
    - 1.6.2 Calibrations 129
    - 1.6.3 Measurements 133
    - 1.6.4 Discussion on the microheater geometry 136
    - 1.6.5 Thermal uniformity measurements 138
    - 1.6.6 Measurements at high temperatures 142
        - Experimental results 142
        - Physical and structural behavior 148
        - Summary 149
    - 1.6.7 Thermal ageing 150
        - Tests at constant current 150
        - Tests at constant voltage 153
        - Physical explanations 156
        - Summary 157
    - 1.6.8 Reliability 158
- 1.7 Conclusions 159

2 Microheater based flow sensor 163
- 2.1 Introduction 163
- 2.2 Design and fabrication 164
- 2.3 Measurements results 168
    - 2.3.1 Calibration and temperature measurements without flow 168
    - 2.3.2 Microheater and thermopiles response time measurements without flow 173
    - 2.3.3 Flow measurements 175
        - Introduction 175
        - Sensor in configuration 1 176
        - Sensor in configuration 2 179
        - Sensor in configuration 3 184
        - Discussions and summary 185

2.4 Discussions and comparison with the state-of-the-art 187
2.5 Conclusions 188

3 Gas Sensors on microhotplate 193
3.1 Introduction 193
3.2 Interdigitated electrodes: from design to deposition 195
    3.2.1 Design 195
    3.2.2 Metal selection and deposition 196
3.3 Sensitive layer deposition 200
3.4 Summary of the fabrication steps 203
3.5 Measurements results without gas 205
3.6 Measurement results under gas and discussions 207
3.7 Conclusions 212

4 SOI-CMOS compatibility validation 213
4.1 Introduction 213
4.2 Basics of SOI technology 215
    4.2.1 The main SOI technology issues 216
    4.2.2 The transconductance over drain ratio current 218
    4.2.3 The properties of the oxide and the Si/SiO_2 interface traps 218
4.3 Post-processing steps 220
4.4 Measurements 220
    4.4.1 MOS transistors 222
        Measurements results 222
        Discussion 225
    4.4.2 MOS capacitors 230
    4.4.3 Gated Diodes 232
    4.4.4 Summary 234
4.5 Transistors on membrane as final demonstrator 235
    4.5.1 Design and fabrication 235
    4.5.2 Measurements and discussion 238
    4.5.3 Applications 241
        Pressure sensor 241
        Gas sensor 242

| 4.6   Conclusions 243

IV   Conclusions and outlook   245

Appendixes   253

A   (100) Silicon crystallography   255

B   About Interferometry...   259

C   About Reflectometry...   263

Bibliography   265

Publications originated from this work   285

Index   289

Part I

Introduction: Context and motivations

Sensors are devices that provide an interface between the electronics and the physical world. When converting non-electrical physical or chemical quantities into electrical signals, the sensors help the electronics to "see", "hear", "smell", "taste", and "touch" [1]. Sensors have become an essential element of process control and measurement systems in almost all spheres of our life and tend to be more and more miniaturized. Electrical signals coming from sensors always need to be processed in order to be understood by the sensor user, and require therefore additional electronics. Unfortunately, sensor fabrication techniques (such as micromachining) are only more or less compatible with electronic circuits which are processed using the well established Complementary Metal Oxide Semiconductor (**CMOS**) process. These circuits can then be placed far away from the sensor and its environment to sense, and can be connected to the sensor by wires of which the length can vary. Sensor and electronics can also be implemented on two different chips connected together with thin wires in the same package, and placed in the environment to sense. In both cases, electronics and sensors are separately processed and connected together at the end of the process so as to avoid compatibility problems during their fabrication. Such solutions are named *hybrid solutions* since they use microelectronics fabrication techniques but cannot be merged due to some incompatibilities. Most of the high production sensors -such as pressure sensors and accelerometers- which we can find in the automotive sector, are processed this way thanks to their lower cost (less processing steps).

Nevertheless, in order to shrink their size, to increase their performances, their reliability and their economical added value, it intuitively seems more efficient to integrate the sensor and its electronics on the same silicon die. This technique is thus called **co-integration** and the resulting co-integrated device is called *smart sensor* or *microsystem*. In such fabrication processes, the compatibility between CMOS techniques and the ones that are necessary to build sensors must be complete.

The purpose of this work, by means of typical sensors realizations, is to add contributions to the growing research in this promising field (see more particularly the works of Baltes and its staff, in [2]). We will especially focus on a specific process technology, Silicon-on-Insulator (**SOI**), which is the best candidate for successful co-integrations in the largest kinds of applications. Only sensors based on **micromachined thin film membranes** will be studied, especially thermal sensors which need to be built on dielectric membranes to ensure a maximal thermal insulation and maintain power as low as possible.

In this introductive part, we firstly justify our choice regarding the fact that all this work is based on SOI technology. We explain then why thin film membranes are essential to successfully sense the surrounding world as well as to increase the performance of some integrated circuits. A debate follows about the advantages and drawbacks of the co-integration on the same chip. Finally, the table of contents is detailed.

0.1   Why Silicon-on-Insulator technology ?

Silicon-on-Insulator (SOI) technology is the most advanced of the present CMOS technologies [3]. This technology offers the possibility of building electronic devices on a thin layer of silicon that is electrically isulated from the thick silicon substrate through the use of a buried dielectric layer. In standard silicon technology (named bulk Si), the silicon substrate is associated with undesirable effects such as high leakage currents from the source and drain towards the substrate, parasitic bipolar components, parasitic source and drain capacitances and more importantly, interference between individual active devices or circuits built in the same integrated chip [3]. The use of Silicon-on-Insulator (SOI) substrate is likely the best way to overcome the limitations of bulk technology. The insulating layer blocks charge transport between the active layer and the substrate, featuring this technology as the best one for radiation and high temperature environments, like outer space [4][5], military or industrial, avionics and automotive sectors [6][7]. In addition, the reduction of parasitic capacitances and leakage currents allow better high-frequency performances and lower power consumption when compared to bulk counterparts [8]. Finally, SOI devices benefit from improved insulation between devices which allows higher device density compared to that possible in bulk Si [4].

The first confirmation that SOI technology has become the state-of-the-art technology in low power high speed ICs came when IBM announced its first fully functional SOI

0.1. WHY SILICON-ON-INSULATOR TECHNOLOGY ?

mainstream microprocessor in 1998. More recently, AMD started its new generation of high-end microprocessors based on SOI technology. Nowadays, SOI ICs are only a fraction more expensive than their counterparts made in bulk technology (10 or 15 % more per die [4]). This is due to the high initial costs of the SOI wafers (3 or 4 times as much as bulk silicon [4]). These are however counterbalanced by the reduction of process steps and by the higher packing density of ICs per wafers. In addition, it is expected that as the volume of SOI wafers increases, the costs will significantly decrease, i.e. making SOI the favorite route to ultimate MOS devices [3][4].

Recently, SOI technology has also been applied to Micro-Electro-Mechanical Sensors (MEMS) [9][10]. The buried oxide present in the SOI material can be very successfully used as an etch-stop layer for both wet and dry *bulk micromachinings*, either from the back or from the front side of the wafer. When releasing the silicon substrate from the backside of the wafer for instance, a stacked buried-oxide/silicon membrane can be obtained. The high quality of the top crystalline silicon film enables semiconductor devices such as temperature sensors, microheaters, MOSFETs, microwave circuits to be built within the membrane in order to get a better thermal or electrical insulation. This membrane can also play the role of support for mechanical sensors to sense pressure for instance, and integrate their sensing elements such as piezoresistors, directly in the silicon film. The use of SOI material is imperative in the cases where a silicon membrane is desired.

The merit of silicon as a mechanical material in comparison to other ones do not need to be reminded. The ability to use Si as a superficial layer for surface micromachined structures is a great benefit and can be fulfilled by SOI technology. In this case, the *surface micromachining* consists in selectively etching the buried oxide, acting as a sacrificial layer, in order to release silicon structures in suspension above the substrate. The sacrificial layer features in this case a high etching selectivity compared to the superficial layer and the substrate (both silicon layers), and a high degree of uniformity. Capacitive structures between the top silicon film and the substrate can be built this way [9].

The use of SOI thereby appears as the best approach for processing either circuits or sensors. In addition, it offers the unique advantage to save a great deal of design time and efforts in fabrication of microsystems, combining the electronics and the sensors on the same chip, in order to improve their thermal, electrical and mechanical performances. Furthermore, the high temperature and radiation hardness of SOI technology make it the best option to achieve microsystems dedicated to work in harsh environment, such as automotive, aerospace and military sectors.

0.2 Why a thin-film membrane ?

Our work especially focuses on sensors requiring a thin film dielectric membrane to achieve a great **thermal insulation** coupled with a low electrical power consumption. It is well known that silicon conducts heat very well (its thermal conductivity is about 100 times higher than silicon oxide [11]) and therefore leads to consume a lot of power to reach a temperature as high as 400°C. Dielectric membranes, as thin as 1 μm, constitute thereby the best choice for thermal insulation.

A dielectric membrane can also play two main other functions:

- A significant purpose of a dielectric membrane is its ability to provide with **electrical insulation**. A typical application based on this purpose is demonstrated when integrating a meander microwave inductor (Fig. 1(a)) on the membrane and comparing its quality factor as well as its resonance frequency with and without the silicon substrate. The results depicted in Fig. 1(b) show that the membrane enables to increase the resonance frequency as well as the quality factor of the inductance thanks to the loss of parasitic capacitances to the substrate. The same kinds of results were reported in [12].

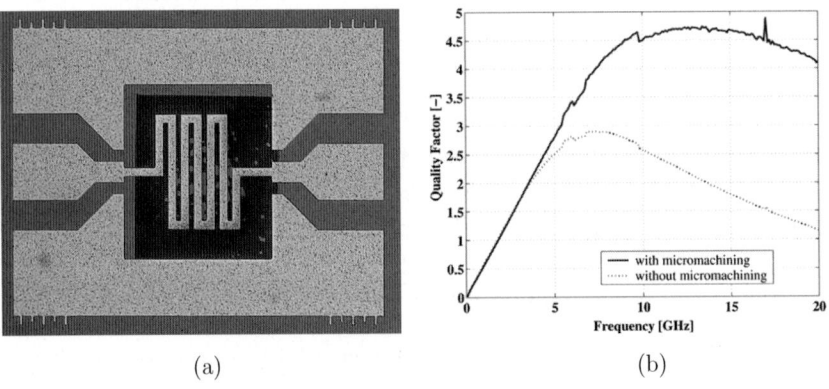

Figure 1: (a) Picture of a meander inductor (0.65 nH) on our dielectric membrane; (b), quality factor Q versus frequency for the meander on membrane and on substrate.

- A dielectric membrane can also play the role of **mechanical support** for mechanical sensors, such as piezoresistive pressure sensors (Fig. 2). In such sensors, the

0.2. WHY A THIN-FILM MEMBRANE ?

membrane is deflected by a difference between the upper and lower applied pressures. The membrane deformation induces a resistance variation in the 4 polysilicon piezoresistors connected in a Wheatstone bridge, and then a voltage difference across the bridge, that is proportional to the differential pressure which is applied [13].

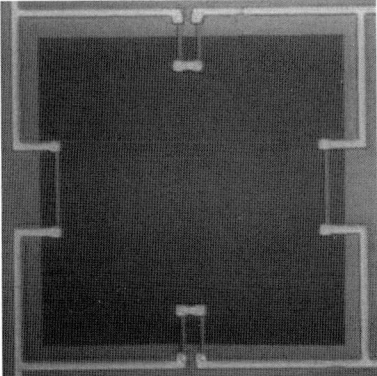

Figure 2: Pressure sensor on dielectric membrane based on 4 polysilicon piezoresistors connected in Wheatstone bridge.

It must be noted that in some cases, the dielectric membrane that is built from the standard inorganic layers of the microelectronics fabrication can be successfully replaced by an other constituting material, i.e. *Polyimide* for instance. This is a new promising organic polymer, fully CMOS compatible, that features high dielectric properties, high planarization, excellent self-adhesion, high thermal stability, ease of patterning (coating by centrifugation and photosensitive), great chemical stability and good mechanical properties such as high elasticity and low thermal dilatation coefficient [14]. In addition, it offers the great advantage to be deposited in much higher thickness compared to oxide or nitride. Polyimide can be successfully used as an electrical insulator for high frequency sensors as reported in [15], or as a thermal insulator for thermal sensors [16]. Furthermore, it can be noted that -thanks to its high water permeability in some formulas- it can constitute the sensing layer of a humidity sensor [17].

0.3 Why co-integration and CMOS compatibility ?

In some applications, it can be sufficient to package the sensor and the signal-processing electronics on two separate chips (or dies), but in a single module, using hybrid technologies [13]. Such technologies thereby require interconnection wires between the sensor-chip and the electronic-chip, processed separately but finally integrated in the same package. This technology allows for performing very low cost (but low performance) sensors on their own chip, coupled with powerful electronics dedicated to offset their defects, on a second chip in the same package. This technology is well established in many high productivity sensors companies due to its low cost per chip since requiring less processing steps.

On the other hand, the co-integration on the same chip, of the signal circuitry with the sensor structure is determined mostly by how the performance, the reliability, the size gain, the overall yield, the packaging and the overall system cost are affected. Nevertheless, in many applications, to reach high performances, co-integration nowadays appears inescapable and can plentifully be justified by an eventually lower starting cost.

In the first example (electrical insulation) outlined here above, we showed that the bulk micromachining, which is a typical sensor-fabrication technology, and passive microwave circuits require to be intimately co-integrated to increase their frequency performance. In addition, when the typical described inductor needs to be integrated into a more complex circuit (in a high-frequency and low-power oscillator for instance), the hybrid technology is unthinkable due to the too high parasitic capacitances between the interconnections wires which would increase the response time, as well as the power consumption.

On the contrary, the pressure sensor of the second example does not need to be especially co-integrated with its electronics on the same chip to be impressive. Nevertheless, polysilicon piezoresistors requiring a Wheatstone bridge to provide an interpretable signal, are more and more replaced by MOS transistors in order to make an easier signal processing and to get an increase in the sensor sensitivity in some applications [18]. An example will be discussed in the last chapter (last part). In this case, co-integration still appears to be an asset.

Regarding gas sensors, solid state gas sensors of semiconductor type deposited on ceramic substrates have been known for more than 25 years, but it is only recently that the possibility of using microelectronics technology to fabricate them has been contemplated [19]. This technological revolution offers several advantages such as small size, low

0.3. WHY CO-INTEGRATION AND CMOS COMPATIBILITY ?

weight, low power consumption, more uniform temperature distribution, low cost due to automatic and batch production [19] for this kind of sensors. The considerable interest for such sensors arises from the need to find solutions everywhere -from the industrial, agricultural and military activities to the domestic ones- to reinforce environmental regulations for the next decades [20], and fully justifies the growing research in this field. In addition, the possibility to co-integrate the sensor and its electronics on the same chip increases its fabrication easiness, especially in the SOI technology, as discussed above [3]. Consequently, such gas sensing microsystems can be used in radiation or high temperature environments, without affecting their reliability. It would not be the case if the same sensor was integrated in hybrid technology.

Finally, the co-integration interest is to a large extent dependent on economic considerations [13]. An important economic factor is the **yield** of both the sensor and the circuitry. For most semiconductor sensors, it is expected that the yield will be dominated by gross defects, or parametric processing problems (e.g. errors of proportions in the preparation of a wet etchant), as opposed to the situation in integrated circuits where random defects (e.g. pinholes, photolithographic defects) dominate the yield. This is due to the difference of area between circuits and sensors, and their difference of physical and structural complexity. The yield of integrated circuits is strongly dependent on the area of the circuit. Therefore, while the circuit yield is determined especially by the circuit area and the defect density, the overall yield of a co-integrated sensor-circuit chip will moreover depend on the number of additional processing steps that have to be performed, which results in its reduction [13].

An other important problem linked to sensors is their **packaging**. No sensor can be designed without a consideration of the final package and the influence on its final costs [13]. The package for each type of sensor is different and has different requirements as it must transfer the variable being measured to the sensor while protecting the sensor from harmful environmental effects [13]. When sensors and circuits are co-integrated on the same chip, the packaging becomes easier and cheaper to design and fabricate, especially thanks to the reduction in the number of wire bonds. Consequently, the long-term reliability is increased as well as the miniaturization.

In conclusion, in most cases, co-integration may increase the costs because of the additional processing (and therefore decreased yield) and design efforts of the individual Si devices. However, the manufacture final cost of a reliable system is not always increased. Packaging problems are reduced, co-integration increases the long-term reliability as well as the miniaturization and it can improve the overall system performance with less components, which results in a higher quality system. Furthermore, if SOI technology is used, the microsystem performances as well as its design easiness are increased, featuring finally high performance sensors which often offset the higher costs of the starting SOI material.

0.4 Contents of the work

The content of the present work is divided in two main sections following the introductive part I. Part II will focus on the techniques and materials, especially the TMAH silicon bulk micromachining used to release our membranes and the measurements of the residual stresses in the thin dielectric films which will constitute our membrane. Part III will describe our sensors realizations based on such thin dielectric micromachined membranes. Two sensors will be investigated, both of them based on a unique basic cell, a microhotplate designed to uniformly heat a given area of the membrane. This last part will be concluded by an in-depth study aimed at validating the fully SOI-CMOS compatibility of our devices.

CHAPTER 1 of PART II is focused on a particular silicon micromachining technique based on the TetraMethyl Ammonium Hydroxide (TMAH) wet etchant dedicated to machine a dielectric membrane from the silicon substrate. This well-known technique has been recently introduced in the world of MEMS. Tabata et all. [21] reported in 1992 their first tests with this promising technique. This etchant featured the great advantage, in comparison with former ones, to be fully CMOS compatible and easy to handle. The amount of publications on this technique has slightly been increasing after '92 with a first significant peak around 1998. Nevertheless, a lot of review papers and books still reported the same results. Nowadays, this technique is well documented but needs to be carefully reviewed in order to be well defined. Our review on the TMAH etching is the first original contribution of this work. All its etching properties are described and illustrated with our own tests performed with our in-house optimized etching benches. The selectivity of TMAH versus aluminum is especially studied and we develop for the first time a new etching method, based on new proportions of the needed additive chemicals, for allowing the etching of a whole processed wafer fully immersed in the etchant. Furthermore, the selectivity of TMAH versus other metals is tested and reported for the first time. Finally, a particular etching property of TMAH is developped, i.e. its ability to etch under a masking material in order to build suspended microstructures. This property is plenti-

0.4. CONTENTS OF THE WORK

fully illustrated by the fabrication of test beams, bridges and rings aiming to extract the residual stress in thin dielectric films. This is the purpose of the second chapter.

CHAPTER 2 of PART II is relative to the stresses which appear in the films during their processing and which remain in residual-stress form in the processed mechanical structures. Particularly, residual stresses (compressive or tensile) affect the profile as well as the robustness of a thin dielectric membrane. Various methods have been published to extract the residual stress in polysilicon and in nitride layers. Nevertheless, residual stresses in silicon oxide have not been much studied so far. Stress measurements of combinations of oxide and nitride layers are even more rare. So, the major original contribution of this chapter is to detail and compare two measurement methods of the residual stresses in the thin dielectric films constituting our membrane, either when they are separated or when they are combined. The first method is the well-known substrate curvature technique, here generalized to extract the residual stress in thin-stacked films. The second one is based on micromachined microstructures newly designed for our purpose and to be released using the appropriate underetching property of TMAH. Finally, the residual stress as well as the variation of stress throughout the membrane thickness have been extracted by the two methods and compared. This technique appears to be a new way to design robust membranes with well-controlled residual stresses.

In PART III, CHAPTER 1 deals with the design of a new low-power microhotplate. To decrease the power consumption, it is well known that this kind of device needs to be insulated from the substrate, on a thin dielectric membrane for instance. Our strong membrane optimized in chapter 2 of part II constitutes the mechanical support of our microhotplate. The main purpose of our microhotplate is to heat a gas sensitive layer which can only react with ambient gases at temperatures around 300-400°C. Numerous high-performance microhotplate-based gas sensors are reported in literature and many of them are termed "CMOS compatible". Nevertheless, most of them use layers which are not inherent in a standard IC fabrication or use technological steps which are far too complicated to be co-integrated with CMOS circuits. Furthermore, only a few have been actually produced in standard CMOS processes, based on the layers available in such processes. We propose a new co-integrated microhotplate-based gas sensor, that is cheap and which can easily be processed in the frame of CMOS-SOI technology. The advantages of SOI technology on the standard bulk one are clearly shown in this application. The originality of our design also lies in the novel loop shape polysilicon microheater optimized to uniformely heat a given area. The thermal uniformity is measured by means of a thermoreflectometry technique developed for this purpose. Polysilicon features well know drawbacks regarding its thermal stability with time when exposed at high temperatures (from 400°C) but needs to be heated up to 700°C to guarantee a great stability of the gas sensors at their operating temperatures. An other originality of our work is to study in depth the behaviour of the polysilicon resistor at high temperature as well as a function of time. Some works in this field can be found in literature but always suffer from serious deficiencies. A lot of other measurements have opened the discussion on the effect of the membrane size and microheater loop shape on the power consumption. Finally, the limitations of our design are detailed and some improvements proposed for future designs.

CHAPTER 2 of PART III moves on to the expansion of our optimized microhotplate towards calorimetric flow sensor. A lot of such sensors based on many different sensing principles can be found in literature. A complete review of most recently published flow sensor results is reported and reveals that many of them are still referred as "CMOS compatible" but use materials which are not available in standard low cost IC processes. Our calorimetric sensor is based on a microheater (on membrane) integrated between two thermopiles (i.e. integrated thermocouples) to sense the temperature variations when a flow passes on the surface of the sensor. Three configurations are compared and help us to complete our thermal study about the microheater as well as about the membrane for a better understanding of their behaviour. Finally, the results of our various measurements under flow reveal the main characteristics as well as the limitations of our design.

CHAPTER 3 of PART III develops the work that had been started in the first chapter of this part, in order to expand the microhotplate towards a gas sensor. The choice of the metal needed to build the interdigitated electrodes, designed to sense the gas sensitive layer, is carefully studied, followed by a description of the gas sensitive layers deposition. The chapter concludes on some measurements with and without gas.

CHAPTER 4 of PART III validates the SOI-CMOS compatibility of our process. The chapter starts with a helpful summary on SOI technology, its technological issues and a short section about oxide charges and interface traps. Several integrated devices such as transistors of different types and sizes, capacitors and gated diodes have been measured after each critical step of our post-process. The impact of each post-processing step on the integrated devices has then been carefully discussed comparing electrical characteristics such as the transconductance over drain current ratio (versus drain current normalized to width/length aspect ratio), the recombination currents, the leakage currents through the gate, ... The conclusions of our measurements constitute a really original contribution since the literature never in depth reports the interactions between CMOS circuits characteristics and post-processing. A final demonstrator consisting in n- and pMOS transistors located on a dielectric membrane has been studied. It illustrates and demonstrates in one device the full SOI-CMOS compatibility of our process and opens the door to really interesting new developments, such as high performance pressure sensors, gasFET or thermodiodes-based flow sensors.

Finally, a CONCLUSION in PART IV summarizes this work and introduces future research perspectives.

Part II

Techniques and materials

Chapter 1

Silicon bulk micromachining with TMAH

1.1 Introduction

A classical approach to the fabrication of integrated sensors is to use standard (or lightly modified) integrated circuits (IC) processes, and to enhance it with one, or several, post-processing steps, as micromachining [22]. This method has the benefit that it enables merging IC circuits and micromachined structures with high performance on the same chip. It needs nevertheless to trade off flexibility in the sensor structure design and ability to easily fabricate fully integrated systems on one chip. Instead of radically changing the circuit processing to conform it to the sensors fabrication steps, the post-processing should be better compatible with the integrated circuits process to allow their mass production [22][23].

For example, typical integrated gas sensors or integrated flow sensors require the circuit fabrication in first step, followed in final step by the back side micromachining of the wafer to create the thermally isolated dielectric membranes. In such sensors case, the membrane supports a polysilicon microheater, an interlevel densified PECVD oxide layer and the metallic wires connected to the circuits located outside of the membrane. Therefore, to etch membranes at the end without damaging standard CMOS integrated circuits, the four following conditions must be respected:

1. Etching must not damage aluminum contacts and densified PECVD silicon oxide on top side;

2. Thermal silicon oxide must be usable to stop the etching on backside;

3. A good selectivity versus silicon nitride is a plus to use it as masking material on back side;

4. Etching must not contaminate CMOS circuits by introducing alkaline ions.

A good selectivity to aluminum, silicon oxide and silicon nitride is therefore required as well as the lack of K or Na ions. Finally, it is better to choose a safe and easy-to-use solution. When further speaking about IC-CMOS compatibility, if all of these conditions are observed, a fully processed wafer can then be directly immersed, with no special measures, in the silicon etchant. In this case, frontside (circuits side) protection is not necessary when etching the backside of the wafer and batches of wafers can therefore be processed in same time. Furthermore, it is highly practical to be able to use a fully compatible CMOS etchant when etching on front side, i.e. on the same face as the circuits. This technique can even be used on packaged and bonded circuits as demonstrated in [24]. We will finally see further that a good selectivity to aluminum allows its use as a high quality mask to protect silicon from etching instead of other layers more difficult to deposit in post processing.

If such compatibility is not provided, a mechanical holder is unavoidable to protect the front side of the wafer since no other front side protection coating is sufficiently reliable. It is particularly the case when polymer (as polyimide for humidity sensors) or screen-printed tin oxide (for some gas sensors) is deposited on front side at the end of the process, just prior to the back side etching. In this method, the wafer is held in a holder (often made from PEEK) in order to hermetically seal the front side from the etchant solution, allowing most often two wafers to be processed at the same time. The wafer is fixed between O-rings which are carefully machined in order to avoid mechanical stress in the wafer (Fig. 1.4). Furthermore, the process reliability is increased by a venting hole that avoids pressure to build up in the cavity behind the wafer when the closed holder is transferred into the hot etchant. However, if a membrane breaks before the end of the etching, etchant leaks on circuits side and damage them.

So, in the following chapter, some generalities about silicon micromachining will be firstly reminded including a comparison between most silicon etchants. Only one solution observing the previous CMOS compatibility conditions will be chosen, the anisotropic tetramethylammonium hydroxyde etchant or TMAH. A description of TMAH etching will follow with its specific introduction in our clean rooms and its properties, illustrated with our experiments. Its selectivity versus dielectric, aluminum and other metals will then constitute three special sections. And we will finish with the two main properties of TMAH: its capability to stop its etching on some layers and its undercutting inherent to the etchant anisotropy.

1.2 Generalities about silicon micromachining

The purpose of silicon *bulk micromachining* is to selectively and locally remove significant amounts of silicon from a silicon substrate [25]. In comparison with s*urface micromachining* achieved by building up starting from the substrate surface, the bulk micromachining is performed by digging in the substrate [26]. Only bulk micromachining will will be of interest in the present context.

Not only membranes but also a wide variety of structures as holes, bridges, cantilevers can be fabricated using bulk micromachining or etching of silicon. Etching can be done either in liquid form (*wet etching*) or in gaseous form (*dry etching or plasma deep reactive ion etching - DRIE*). Although deep RIE has become popular for realizing high aspect ratio silicon microstructures, the advantages of wet etching technology such as low process cost, better surface smoothness and lower environmental pollution make them a complementary technology [21]. Furthermore, wet etching features an effective etch stop in contact with dielectrics or other layers. We will focus only on wet etching in this work. Wet silicon etchants can be divided into *isotropic* and *anisotropic* types [22]. Isotropic etchants etch in all directions at the same rate whereas anisotropic etchants etch much faster in one direction than another [11]. A typical isotropic etchant is a combination of hydrofluoric acid (HF), nitric acid (HNO_3) and acetic acid (CH_3COOH), also known as HNA [22]. It leads to rounded etched structures (as on Fig. 1.1) which geometry is dependent on the agitation of the etchant. The etch can be masked with silicon nitride or silicon dioxide but the latter is attacked fairly quickly [25]. Due to the lateral undercutting and resulting lack of dimensional control and reproducibility, isotropic etchants are not often used in micromachining [22] in spite of their very fast etching rates at ambient temperature.

Anisotropic etchants shape or "machine" desired structures in crystalline silicon and are more powerful but need for etching temperature higher than the room temperature (typically 80-90°C). Anisotropic etching results in geometric shapes bounded by perfectly defined crystallographic planes[1] [27]. As explained in Appendix A, a (100) silicon wafer

---
[1]See Appendix A for a summary about the repartition of the crystallographic planes in a oriented

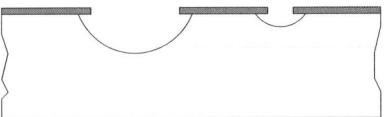

Figure 1.1: Cross section of an isotropic etched cavity.

has a surface coplanar with the (100) crystal plane. In anisotropic etchants, the (100) plane etches much faster than the (111) crystal plane oriented at 54.74° to it [22][25]. As illustrated in Fig. 1.2(a) a square mask opening will result in an inverted pyramidal cavity, truncated or not depending on the width of the mask and the etching depth. In case (a), the square mask aligned with the [110] direction has only concave corners and the etching is stopped at {111} intersections [11]. But if the same square concave mask is misaligned with the [110] direction, undercutting will occur (Fig. 1.2(b)). As a general rule, assuming a low etch rate of {111} planes, a mask opening with arbitrary closed geometry and orientation (as shown on Fig. 1.3(a)) which is exposed to an anisotropic etchant for a sufficiently long time will produce an inverted pyramidal cavity, the base of which is determined by the smallest rectangular shape that contains the entire pattern [22][28]. Furthermore, if convex mask corners are exposed to anisotropic etchant (Fig 1.3(b)), they become undercut along other crystal planes and the etchant tends to circumscribe the mask opening with {111} walled cavities [27]. We will later come back on these undercutting behaviors.

There are several anisotropic silicon etchants, such as EDP (ethylene diamine, pyrocatechol and water), KOH (potassium hydroxide) and TMAH (tetramethylammonium hydroxide). Their principal characteristics are listed in Table 1.1 [27][11][25][1][21]. These anisotropic etchants are more or less selective to dielectrics such as silicon oxide and silicon nitride and to metal as aluminum. Anisotropic etching can also be stopped on heavy boron doped silicon junction (p^{++} etch stop) or at biased P-N junctions (electrochemical etch stop). Boron doses resulting in a decreasing of the silicon etch rate are included in Table 1.1.

Each etchant has its advantages and problems. The well known KOH etchant provides the best selectivity for {111} planes versus {100} planes to produce well defined and controlled cavities and very smooth etched surfaces. But it is not fully CMOS compatible as it contains alkali ions (potassium) which can introduce charges under MOS transistor

---

(100) silicon wafer, the standard wafer used in CMOS processes.

1.2. GENERALITIES ABOUT SILICON MICROMACHINING

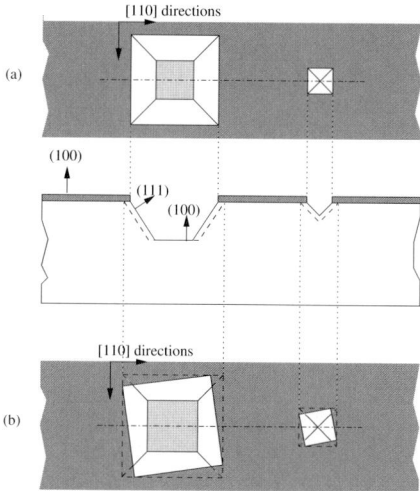

Figure 1.2: Illustration of anisotropic etching of (100) silicon. Cross section on the middle; (a) top view when the mask is perfectly aligned with [110] direction; and (b), top view when edges of the opening are misaligned from [110] directions resulting in undercuting as shown in dotted lines on cross section.

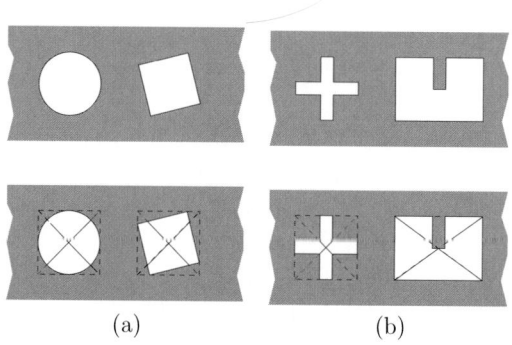

Figure 1.3: Resulting structures due to undercutting of different mask opening; (a) misaligned structures and (b) undercutting at convex corners.

Table 1.1: Table comparing the general features of various anisotropic etchants suitable for bulk micromachining applications.

	EDP	KOH	TMAH
Si etch rate [μm/min]	0.75 to 1.25	1 to 2	≈ 1
SiO$_2$ etch rate [nm/min]	1 to 80	1 to 10	0.05 to 0.25
Si$_3$N$_4$ etch rate [nm/min]	0.1	very low	0.05 to 0.25
(100)/(111) etch ratio	35	400	10 to 35
Si roughness	low	very low	moderate
Al selectivity	no[1]	no	yes[2]
Au selectivity	yes	yes	yes
p^{++} etch stop [cm^{-3}]	B>7x10^{19} $\hookrightarrow$ER/50[3]	B>10^{20} $\hookrightarrow$ER/20	B>2x10^{20} $\hookrightarrow$ER/40
Alkali ions ?	no	yes	no
Cost	high	low	moderate
Disposal	easy	difficult	moderate
Safety	low	moderate	high

(1) One specific fomulation does not attack Al
(2) If specific amounts of silicon are dissolved in the etching solution
(3) Silicon etch rate divided by 50 if Boron doses higher than 7x10^{19}cm^{-3}

gates and cause threshold voltage shifts [22], corrodes aluminum and is not so selective versus silicon dioxide. EDP would be more attractive as it does not contain sodium or potassium ions and has a high selectivity versus dielectrics. But this etchant is not aluminum selective (except in one specific formulation) and it has a serious other drawback: it is toxic and mutagenic [22] and very difficult to handle. So in our case where CMOS compatibility as well the low toxicity and the ease of handling are the main concerns, TMAH seems to be the best choice. It is well known in most clean rooms because TMAH is already used in IC process as the developing solution of positive photoresist [21]. Its silicon etch rate is fair and it features a great selectivity versus dilectrics. It is furthermore CMOS compatible as it does not incorporate alkali ions and since, if additives are dissolved in the etching solution, aluminum is passivated and can therefore be preserved [22]. Its drawbacks are a lower selectivity versus {111} planes and a surface morphology rougher than obtained with the other common etchants [11]. But it is not so critical in our cases of study.

1.3 TMAH silicon etching

The mechanism of TMAH etching follows alkaline silicon etching and can be summarized into three separate steps [29][25][1][30][31]

$$(CH_3)_4 NOH \Longrightarrow (CH_3)_4 N^+ + OH^- \quad (1.1)$$
$$Si + 2OH^- \Longrightarrow Si(OH)_2^{2+} + 4e^- \quad (1.2)$$
$$Si(OH)_2^{2+} + 4H_2O + 4e^- \Longrightarrow Si(OH)_6^{2-} + 2H_2 \quad (1.3)$$

In a first step (Eq. 1.1), TMAH is reduced to form hydroxyl ions. In the second step (Eq. 1.2), the silicon atoms at surface react with these hydroxyl ions to form oxidized silicates $Si(OH)_2^{2+}$ and four electrons are injected from each silicon atom into the conduction band. Simultaneously, water is reduced to provide more hydroxyl ions which are bonded to the silicate formed in second step. This final reaction (Eq. 1.3) produces soluble silicic acid, with hydrogen gas as a byproduct.

As shown by the previous equations, the more water will be present, the more silicon will be etched. It was widely reported in [21] and [29] that the silicon etch rate in TMAH decreases when increasing the concentration from 4% to 25%. Below 4%, ions are missing and the etch rate decreases [29]. It is also reported in [1] that temperature increase leads to higher silicon etch rates. But in practice, etch temperatures of 85-90° are used to avoid solvent evaporation and temperature gradient in the solution [27]. For reproducible results, it is therefore critical to provide a good control of concentration and temperature of the solution.

Silicon etch rate measurements were performed in our lab for various TMAH concentrations from 5 to 25 % at 85°C and 95°C on P type (20-25 Ωcm) oriented (100) wafers. Note that it is critical to precise the type and doping level of the silicon to etch since they have an impact on the etching as reported in [32] (it is not surprising since the etching process is fundamentally a charge-transfer mechanism as shown in Eq. 1.2 [28]). The etchant concentration was kept constant by using a water cooled reflux condenser on top of the TMAH beaker to allow evaporated water to condense and drip back into the solution. The temperature was controlled using a PT100 probe immersed in the solution (through a quartz tube to avoid damaging the proble platinum material) or sticked on the quartz bath and connected to a temperature controller monitoring the hotplate (Fig. 1.4(a)). Thermal oxide was patterned with squares to be used as mask. Each test using new samples and fresh solution was performed during 1 hour, after 15 seconds dip in 2

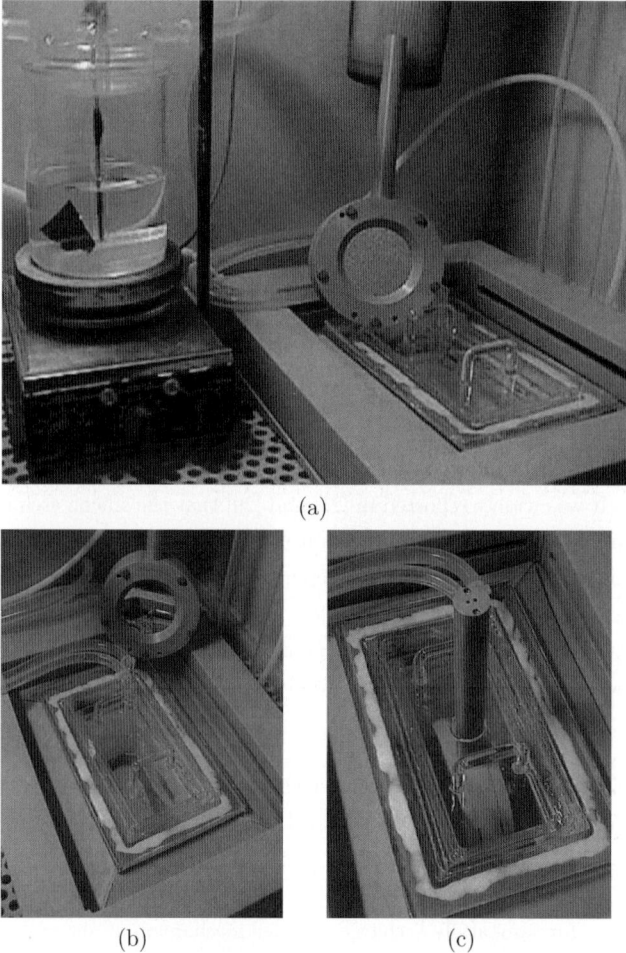

Figure 1.4: Pictures of our etching facilities: (a) experimental becher for samples etching on the left and process bath with the mechanical holder on the right; (b) close-view of the process bath and the holder; (c), close view of the process bath with the holder in position. The reflux condenser can be observed on the bath as well as on the becher.

1.3. TMAH SILICON ETCHING

% HF to remove native oxide on silicon. The volume of etchant was also chosen constant from one etching to an other for a more accurate comparison. It is indeed reported in [32] that a low ratio of etchant volume with respect to silicon exposed area increases its surface deterioration and that is therefore better to use etchant volume as large as possible. All etchings were carried out without stirring (i.e. without magnetic stirrer) in a 1 l quartz beaker (Fig. 1.4(a)). The samples were put in the heating solution from the ambient temperature without special care (on the contrary of technique reported in [33] consisting to heat the sample at the temperature of the solution before its immersion to avoid a transient temperature change). The samples were held vertically in the solution in a dedicated Teflon stand. The etching was performed in ambient lighting conditions since it was reported in [32] that illumination does not affect the etch rate.

The depth of the etched pit was measured with a Dektak profilometer. The results are shown in Fig. 1.5 and agree with published data [1][29][22][34]. Both curves are congruent leading to the conclusion that a same etch-rate can be reached by changing the temperature and the concentration. But other factors such as surface quality and anisotropy (see further) can play a role in the choice of the couple temperature-concentration.

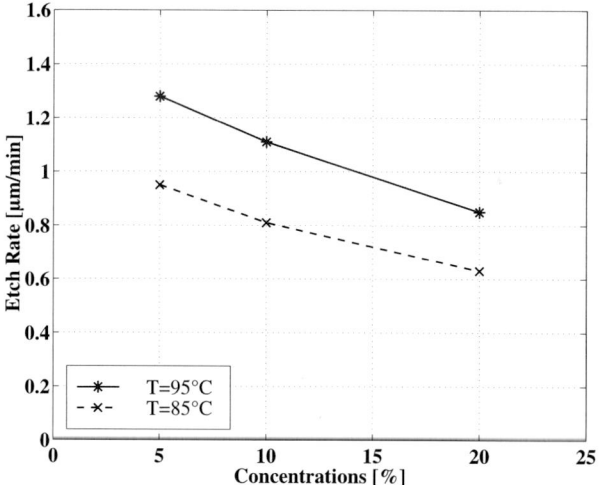

Figure 1.5: Silicon etch-rate of TMAH in various concentrations.

This graphic is essential since there does not exist equations yet to predict etching performances from user-controllable factors such as temperature and etchant concentration. In practice, one typically needs to extract etch rates and characteristics from the literature and by experiment [25]. Consequently, etching time can be monitored to foresee the depth of a cavity. But as we will see further in this chapter, other etch stop methods are more effective.

It can also be shown from Eq. 1.3 that hydrogen is generated on the (100) silicon surface as a result of the etching reaction. This hydrogen can form bubbles which can induce a natural upward flow into the solution causing some little difference in etching rate as demonstrated in [33]. But bubbles can especially cause local micromasking, resulting in hillocks on the etched surface [11]. The nature of the etching mechanism is a competition between the rate of dissolution of silicon and the passivation of the surface by the accumulation of the reaction product [35]. When the concentration of water increases, more hydrogen and therefore more bubbles are present. They locally block the free diffusion of reactants and products to and from the surface leading to the formation of an insoluble precipitate at the point of contact between the bubble and the silicon. This temporary mask will be slowly dissolved upon release of the hydrogen bubble but the pyramid was already grown [29][36]. Initially the number of pyramids at the surface is small, but as etching continues, new pyramids are formed and begin to superimpose over previous hillocks [22]. Therefore, roughness increases with etching time and reduces progressively the etch rate in $\langle 100 \rangle$ directions because pyramids expose slowly etching crystal $\{111\}$ planes (see [35] and [37] for more details about the formation and transformation of hillocks during the TMAH etching). Hillocks are finally unaesthetic. It is therefore critical to eliminate hillocks formation in most cases.

Roughness decreases when TMAH concentration increases since the amount of water present in the solution decreases in same time and therefore produces less hydrogen. It is reported that pyramids vanish with concentrations of TMAH higher than 22% [21]. It is also reported in [38] that roughness increases slowly when increasing the etchant temperature. To prevent hillocking, the best method is therefore to use more concentrated TMAH at lower temperature. But it results in very low etch rate unusable in most applications. It could finally be damageable for the surface quality since the roughness increases with the etching time. Furthermore, high TMAH concentrations are unavoidable when a high selectivity versus aluminum is required as we will see in next parts. The problem can be solved by adding oxidizers into the solution to consume hydrogen during etching but we will come back later on that. Another way reported in [11] is to use ultrasonic or megasonic agitation to avoid bubbling but it seems to reduce the hillocks in number while increasing their size [39]. Furthermore, this agitation is difficult to carry out and could damage fragile structures.

1.3. TMAH SILICON ETCHING

Table 1.2: Comparison of silicon roughness resulting on TMAH etching at 2 representative concentrations after 1 hour at 90°C.

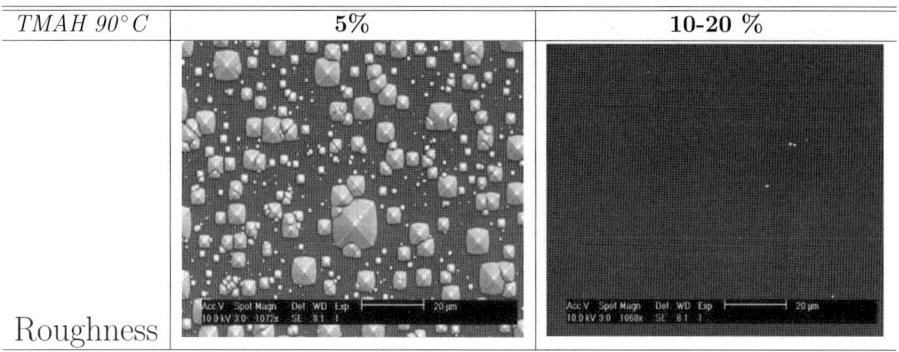

TMAH 90°C	5%	10-20 %
Roughness		

Roughness examinations were performed on samples previously used for etch rate measurements, i.e. after etching time of 1 hour. Table 1.2 summarizes the results obtained at 90°C for various concentrations. Results are depicted as SEM images on a representative window of about 100x100 μm^2. Roughness is difficult to quantify with precision since the number of pyramids, their height and width must be taken into account in same time (an interesting method is proposed in [35] where the number and the size of the pyramids were observed by SEM and quantified using image processing software). Nevertheless, AFM measurements such as depicted in Fig. 1.6 allowed us to quantify the root mean square roughness, decreasing from 64 Å to 8.8 Å when concentration increases respectively from 5 % to 20 % at 80°C. The roughness is confirmed to be decreasing with the concentration and become negligible from concentrations higher than 20 %.

Finally, it is important to note that TMAH is very sensitive to organic contamination, especially at very low concentrations [30]. Like every organic solution, TMAH is also sensitive to rough temperature variations [29]. It is also interesting to note how to clean wafers after the etching. [29] recommends to dip wafers in boiling de-ionised water to stop the chemical reaction and to dissolve any products of the reaction. Cold de-ionised water would not be sufficient to dissolve the precipitates. Finally, methanol is used to dry the wafers instead of rinser dryer too harsh for fragile diaphragms.

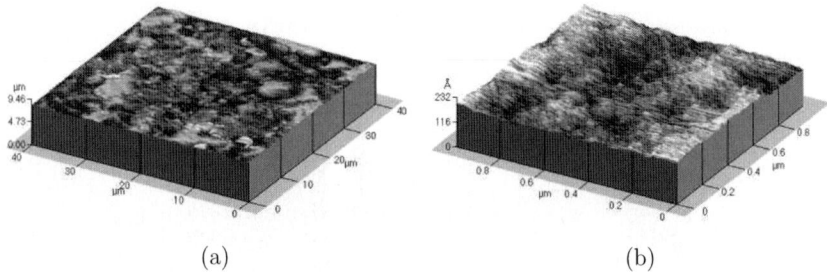

Figure 1.6: AFM pictures of etched silicon with various solutions at 80°C; (a) TMAH 5% and (b) TMAH 20%. A scan of 40x40 μm^2 was achieved for rough surface (a) but for smoother surfaces (b), the scan dimensions were reduced to 1x1 μm^2 to reduce the influence of noise.

1.4 Selectivity versus dielectrics

TMAH exhibits a great selectivity versus dielectrics such as thermal oxide, LPCVD nitride and also densified PECVD oxide. Each of these materials can therefore be used as masking layer to replace photoresist which dissolves immediately in TMAH. It is reported in [11] that silicon dioxide as well as silicon nitride exhibit typical etch rates in the range of 0.05 to 0.25 nm/min. But since it is very difficult to compare grown and deposited process dependent films from one laboratory to another, it appeared essential to characterize the selectivity of these three films made in our lab.

It is reported in [40] and [32] that higher temperature can result in a faster etching of the dielectrics layers; while their etching rate decreases consistently with an increased TMAH concentration. In fact, the trends of the etching rates of dielectric layers are similar to that of the silicon at a difference of a few orders of magnitude.

The etching rates of thermal oxide, LPCVD nitride and densified PECVD oxide were determined by ellipsometric measurements before and after 1 hour of etching at 85 and 90°C for various TMAH concentrations. Selectivities versus silicon were calculated for each concentration by dividing the silicon etch rate by the measured dielectric etch rate. Results are summarized in Fig. 1.7 and reveal that LPCVD nitride shows a better selectivity versus silicon than the other dielectrics.

Note that the thermal silicon dioxide characterized here was a combination of wet and dry growths (68 minutes of wet between two times 20 minutes of dry growth). It is indeed reported in [32] that wet oxide generally etches faster than dry oxide due to its lower quality.

1.5. SELECTIVITY VERSUS ALUMINUM

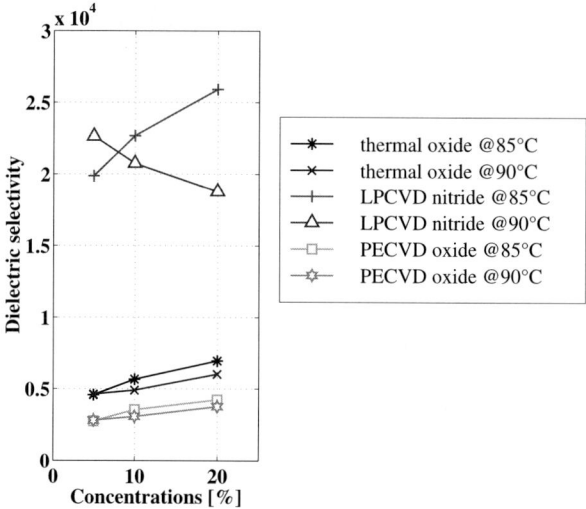

Figure 1.7: Selectivities of various TMAH concentrations versus dielectrics.

Atmospheric pressure chemical vapor deposited oxide (APCVD or pyrox oxide) could also be attractive as protective mask thanks to its high deposition rate (around 0.1 μm/min). Unfortunately, it exhibits pinholes (i.e. defects) and is etched much faster than thermal oxide (as confirmed in [27]). Etching tests were performed with APCVD passivation layer deposited on aluminum to protect it. We observed that TMAH swept through the pinholes and attacked the underlying aluminum, fissuring the passivation layer and unsticking lower layers. Annealing of APCVD oxide may partially remove the pinholes but without decreasing its etch-rate.

1.5 Selectivity versus aluminum

1.5.1 State of the art

TMAH corrodes aluminum as any strong base. A thin layer of aluminum oxide grows on aluminum when it is exposed to air. This oxide layer contains Al(OH)$_3$ which, when exposed to alkaline environments as TMAH etchant, reacts to form alumina by the following reaction [22],

$$Al\left(OH\right)_3 + OH^- \longleftrightarrow Al\left(OH\right)_4^- \qquad (1.4)$$

When the pH of the solution is lowered, the chemical equilibrium shifts toward the left side of reaction (1.4), and there exists a window within which the aluminum is passivated while the silicon still etches [24].

For complete CMOS compatibility, 200 μm thick processed wafers must be etched on back side through the entire thickness during more than 3 hours without damaging aluminum wires on front side. As explained above, protection of aluminum by APCVD oxide was proved inefficient. Furthermore, it is difficult to remove this layer at the end of the process without damaging the other oxide films (e.g. on backside of the membrane). Polymers such as polyimide used as passivation layer was also revealed ineffective. It is preferable to avoid protection by mechanical holder when possible, as explained above. Gold is inert in TMAH but cannot routinely replace aluminum since it is contaminant in clean rooms (it diffuses rapidly into silicon). Other investigated metals, common in microelectronics processes, were also damaged in TMAH (see next section). Decreasing the pH of the solution seems therefore the unique technique to passivate aluminum in TMAH.

It is widely reported that the pH of the solution and therefore the aluminum etch rate can be cleverly reduced by dissolving an appropriate amount of silicon in TMAH [11]. Dissolved silicon in solution forms aluminosilicates on the aluminum which are less soluble than the aluminum oxide that would otherwise form at the metal surface. Introduction of silicates decreases the pH of the solution, slowly lowering the silicon etch rate and works to provide aluminum passivation. As demonstrated by Eq. 1.5, silicates release H$^+$ ions which, by reacting with hydroxyl ions, form more water and therefore decrease the pH [29][1].

$$Si(OH)_4 \Longrightarrow SiO_2(OH)_2^{2-} + 2H^+ \quad (1.5)$$

$$2H^+ + 2OH^- \Longrightarrow 2H_2O \quad (1.6)$$

The silicon can be added in the form of wafer pieces but these take too much time to dissolve. A more suitable method is to add powdered silicon which is available in 99.9 % purity (from e.g. Alfa Aeasar). Experiments reported in [22] showed that the amount of dissolved silicon needed to passivate aluminum increases linearly with the etchant concentration (for a 5 % TMAH, 16 g/l are needed when 35 g/l are necessary for a 10 % TMAH solution). Since it is undesirable to dissolve excessive amounts of silicon in the etching solution, lower TMAH concentrations are preferable. Therefore, a TMAH 5%

1.5. SELECTIVITY VERSUS ALUMINUM

solution is currently reported at a working temperature of 85°C with 16g/l of Si powder to passivate aluminum.

Silicic acid ($Si(OH)_4$) can be added to TMAH as an alternative source of silicates. It is a low density powder and the volume necessary is quite large (e.g. 40g/l for 5% TMAH). Therefore, it is not easy to manipulate and makes the solution opaque. In comparison, the silicon powder has a much higher surface area and dissolves more rapidly. Nevertheless, foaming and excessive bubbling of the solution occurs during its dissolution due to the large amount of hydrogen gas released during this process. To prevent the foaming over the edge of the etch bath, it is necessary to dissolve the silicon at lower temperature (within 40-50°C [41]). After about one hour of dissolution, the solution is quite clear with some not harmful precipitates and can collect the samples. Note that water glass (WG) is proposed in [42] as an other alternative to silicon powder or silicic acid to reduce the time and effort needed for their dissolution but it is not CMOS compatible since it contains sodium ions.

The main drawback is that TMAH solution with low concentration and reduced pH produces more hydrogen since water is added in reaction of Eq. 1.3 due to the dissolution of silicates. The pH decrease leads therefore to rougher etched silicon surfaces leading to a low random silicon etch rate. A strong oxidizer as diammonium peroxydisulfate ($(NH_4)_2S_2O_8$) (also known as ammonium persulfate or APS) can take the place of water in reaction 1.3 to consume hydrogen and thereby improves the silicon surface and the silicon etch rate [29]. [30] reports an increase of the etch rate from 0.77 $\mu m/min$ to 1.02 $\mu m/min$ in a solution of TMAH 5 % at 85°C doped with APS.

Ammonium persulfate is available in powder, it does not contaminate the solution and is CMOS compatible. Moreover, it is safe to use but requires refrigeration at 5-7°C to be preserved [22]. The required amount is only 5 g/l in TMAH 5% [22] but as confirmed in [43], it is rapidly consumed, after about 45 min. in the reaction with silicon. It is therefore needed to add this amount frequently. However, no special care is needed since the addition of excess persulfate is not critical regarding the silicon etching rate [30]. [43] demonstrates that the silicon etch rate is augmented with a corresponding increase in the oxidizer addition frequency. However, the total amount of 5 g/l of added APS every hour is a constant through the literature. Experiments performed in [22] revealed that white precipitates are temporary formed when ammonium persulfate was added in the silicon-doped TMAH solution, due probably to a reaction between the oxidizer and the silicates in the solution.

It is critical to etch with a silicon etch-rate as high as possible since [30] reports that the pH of the solution drops to unacceptably low levels after 6 to 8 hours. The time could even be shorter if large volumes of silicon are etched during use. This decrease in pH is likely due to silicate saturation of the etching solution. But the low cost of the solution and ease of disposal make it feasible to use a fresh solution to finish etching if necessary.

So [22] recommends a solution of TMAH 5 % at 85°C with 16 g/l of silicon powder and 5 g/l of APS. If the concentration is increased to 10 %, 35 g/l are needed and 66 g/l for a TMAH 15 %. A compromise etching condition for the best selectivity between Al/Si, SiO_2/Si and Si(111)/Si(100) was found in [39] at the TMAH concentration of 15% and temperature of 80°C, the Si additive of 60 g/l and the APS additive of 10 g/l. For economical reasons in terms of cost (consumption of TMAH, Si and APS powder) and time, a solution of TMAH 7.5 % at $90°C$ with 45 g/l of silicic acid and 5 g/l of APS was proposed by [43]. A solution of TMAH 10% at 85°C with 60 g/l of silicic acid and 5 g/l is used in [29] in place of TMAH 5 % due to contamination problems. Finally, detailed experiments were performed recently in [41] to optimize the silicon powder and APS concentrations for high etch-rate and high surface quality. They are summarized in Table 1.3. When APS concentrations are taken higher than the upper value of the mentioned range, the silicon etching rate decreases significantly and the silicon surface roughness increases rapidly. From Table 1.3, it appears that the 5 g/l proposed in [29] are not sufficient and could lead to a little damage of the aluminum and a lower silicon etch-rate (however without damaging the silicon surface).

Table 1.3: Results of experiments performed in [41] for two main TMAH concentrations. Note that roughness was measured here as the root-mean-square roughness (over an unknown area).

TMAH $85°C$	5%	10%
+ Si powder	>14 g/l	>32 g/l
+ APS	2-7 g/l	12-20 g/l
Etch-Rate	0.9-1.0 μm/min	0.85-0.9 μm/min
Roughness	< 0.2 μm	< 0.1 μm

1.5.2 Experimental results

Firstly, we used the solution recommended in [22] at higher temperature, i.e. TMAH 5 % at 90°C with 16 g/l of silicon powder and 5 g/l of APS. The amount of powder was validated by the values in Table 1.3. 12 g of silicon powder were first dissolved in 750 ml

1.5. SELECTIVITY VERSUS ALUMINUM

of TMAH 5 %, heated at 50°C to avoid black foam going out the beaker. After around 1 hour, when the solution was quite transparent, 4 g of APS were added, followed by the introduction of the samples. Since APS appeared consumed after 45 minutes, a new amount of APS was added on time and it did not appear necessary to be replenished.

Our tests performed with this solution were more or less successful. Selectivity versus aluminum was perfect as shown on Fig. 1.8 at the end of an etching time of about 3 hours necessary to etch throughout our 200 μm wafer, resulting in etch rates of about 1.2 μm/min (note that an etch rate of 0.8 μm/min was reported in [11] in the same solution at 80°C). In spite of the frequent addition of APS, the surface was not as smooth as expected. This roughness appeared as little silicon pyramids remaining unetched on the backside of the membrane. This is not acceptable in most applications. In some cases etching was never completed, either due to too high hillocks or to any contamination of the solution from some organic compounds. As mentioned above, this problem is more critical at very low concentrations as was the case here. Same problems are reported in [29].

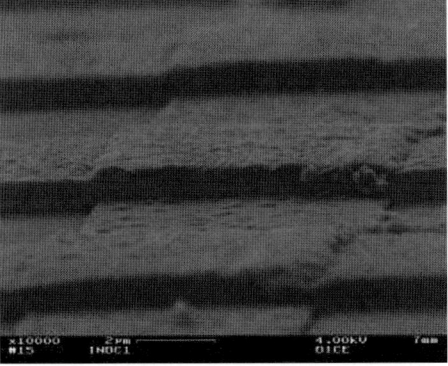

Figure 1.8: Detail of 2 aluminium lines crossing a polysilicon line covered by a silicon oxide layer. The SEM photo has been taken after 3 hours etching in a TMAH based solution (+APS and Si powder). It shows the excellent chemical passivation of aluminium.

Our idea was therefore to compile results exposed in literature to propose a new solution, TMAH 10 % at 90°C with 35 g/l of silicon powder and 15 g/l of APS. In spite of the higher TMAH concentration and the higher consumption of materials, this method yielded a great reproducibility and clean dielectric membranes without silicon residues. Moreover, the same etching time (than with the previous etchant) of about maximum

3 hours was needed to etch completely the 200 μm-thick silicon wafer (i.e. etch rate of around 1.2 μm/min). As for the previous solution, an additional amount of APS (15 g/l) was put into the solution after 50 minutes of etching in order to compensate its consumption.

Table 1.4: Comparison of silicon roughness resulting from Si doped TMAH etching at two concentrations, with and without APS after 1 hour of etching. The insert shows a vertical view of the whole cavity while the larger picture shows a close-view of the bottom of the cavity.

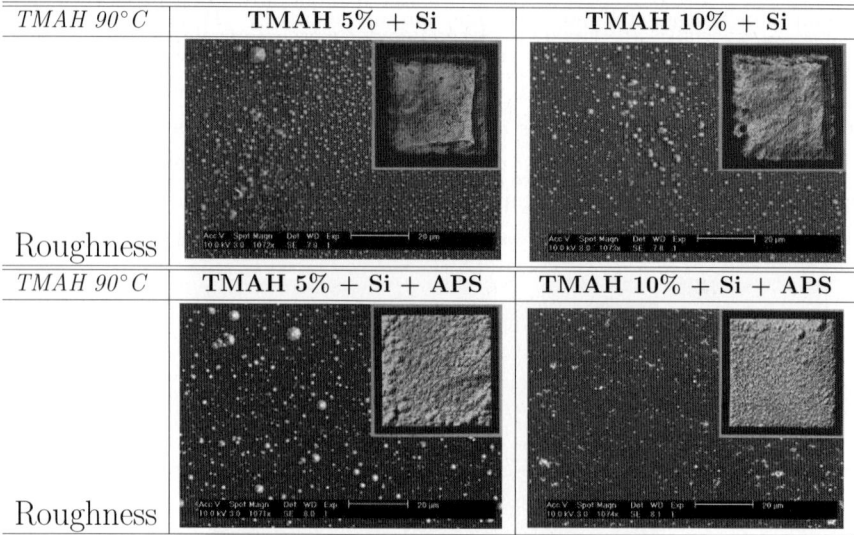

In order to explain these new results, both methods were rigorously compared. Etching was performed on samples in both solutions. After the dissolution of the required amount of silicon (around 1 hour), first samples were immersed in the solution during 1 hour and APS was then added. After 5 minutes of stabilization for APS dissolution, new samples were etched during again one hour. Measurements of silicon etch rate and silicon roughness were performed as previously in each case. Results are summarized in Table 1.4 and Fig. 1.9. First of all, regarding the overall surface planarity on the bottom of the cavity (insert in Table 1.4), it clearly appears that a better etching uniformity is reached when using the TMAH 10 % solution instead of the 5 % one (with or without APS). Furthermore, APS drastically improves the surface quality. When zooming on the bottom of the cavity (large picture in Table 1.4), our results confirm that TMAH 10 % solution gives higher surface quality than TMAH 5 %.

1.5. SELECTIVITY VERSUS ALUMINUM

Regarding the silicon etch-rate, addition of 15 g/l of APS in a TMAH 10 % solution doped with 35 g/l of silicon powder increases the etch-rate. But in a TMAH 5 % solution doped with 16 g/l of silicon powder, the silicon etch-rate is decreased when 5 g/l of APS are added. We observed in fact that APS is consumed faster in the TMAH 5 % solution than in the TMAH 10 %. APS was already fully consumed after 40 minutes of etching in the TMAH 5 % solution while a small amount was still present after 1 hour in TMAH 10 %. It probably explains the silicon etch-rate decrease in our first solution. The selectivity versus dielectrics was observed comparable to the one obtained using undoped TMAH solutions (as confirmed in [43]) and was therefore not measured again. In conclusion, in spite of the higher cost of the solution, our results show that in presence of APS, a higher surface quality can be obtained in TMAH 10 % with a better silicon etch-rate than in TMAH 5 %. Nevertheless, the more severe foaming resulting in the higher amount of silicon powder dissolution needs more care in the preparation of this solution.

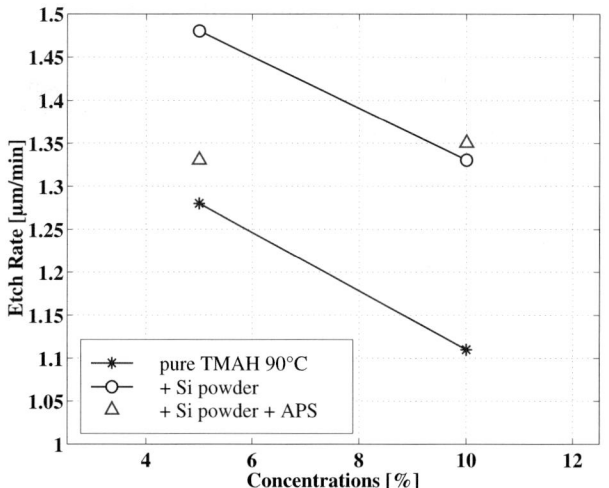

Figure 1.9: Silicon etch-rate comparison in various TMAH solutions measured after 1 hour of etching.

Selectivity versus aluminum was finally controlled in both solutions after 1 hour of etching (in TMAH 90°C with Si powder and APS) and revealed a perfect passivation (etch rate equal to zero) in both cases. Measurements were also performed on sputtered Aluminum-2%Si alloy which in standard CMOS process, advantageously replaces pure aluminum[2]. Its etch rate was also measured equal to zero in both solutions without difference when compare to pure aluminum.

The excellent results extracted from our experiments revealed that our TMAH-based solution constitutes an excellent post-processing etchant presenting reasonable etch rate, high selectivity versus aluminum, silicon dioxide and silicon nitride, safety and ease of use. Consequently, aluminum can even be used as masking material, which can be very practical in a lot of cases. When 500 μm thick SOI wafers are thinned down to 200 μm and polished on backside at the end of a standard IC process for instance, aluminum can be deposited and patterned to be advantageously used as mask for the final backside etching. Note that dip HF performed prior the etching to remove native oxide on silicon slowly attacks aluminum but the etch time is sufficiently short to consider the effect negligible.

An interesting alternative to APS as oxidizer in TMAH-doped solutions is reported in [44]. It consists to add a very small amount of non-ionic surfactant NCW-601A (containing 30% of polyoxyethylene alkyl phenyl ether) into the TMAH solution. It is CMOS compatible and greatly improves the surface quality without influencing the aluminum passivation. Nevertheless, it drastically decreases the (100) silicon etch-rate and can therefore not be used in our case. But it has the great advantage to abruptly decrease the undercutting ratio of convex and concave corners (see further) leading this solution to be very attractive in some applications.

1.5.3 Summary of the TMAH etching steps in presence of Al

Table 1.5 summarizes all of the TMAH etching steps for a typical silicon backside micromachining, in post-processing of a standard CMOS-SOI process. Circuits are present on the top side (with layers such as aluminum and densified PECVD oxide) and aluminum on the bottom side for masking silicon. The etching is performed through the whole 200 μm-thick silicon wafer up to the buried oxide (or thermal oxide when processing on bulk). When the wafers are thicker than 200 μm before this etching step, they need to be thinned and polished to 200 μm.

The TMAH etching is in this case performed in a dedicated rectangular quartz bath covered with a reflux condenser. The temperature was monitored during the etching with

---

[2]Pure aluminum tends to diffuse into silicon and to form aluminum spikes which can make short circuits between thin junctions and substrate. By sputtering Al+2%Si alloy instead of evaporating pure Al, this problem is avoided since aluminum is already saturated in silicon.

Table 1.5: Summary of the TMAH etching steps, from the preparation of the solution to the rinsing of the samples after etching.

	Steps
1	Preparation of the solution (**1.3 L of TMAH 10 %**)
2	Heating of the solution up to **45°C**
3	Addition of **45.5 g** of **silicon powder**
4	Control of the silicon dissolution during $\geq$1h00
5	When the solution is quite, increase of the temperature up to **90°C**
6	When the solution is clear, addition of **19.5 g** of **APS** and mixing
7	Immersion of the wafers into the bath
8	After **50'**, addition of again **19.5 g** of **APS** and mixing
9	After **2h10** of etching, remove the wafers from the bath
10	Immersion of the wafers in a bath of **boiling DI water**
11	Immersion of the wafers in a bath of **cold DI water**
12	Rinsing in **methanol**

a PT100 sensor sticked to the quartz bath (Fig. 1.4). Table 1.5 summarizes the etching steps when the wafers are directly immersed in the etchant. The PEEK mechanical holder can also be used in some cases but is optional if there is no critical film on the front side. In some other cases, we performed the etching in the holder during the 2 first hours and we finished the etching by immersion in the etchant. This technique prevents any leakage of the solution through a possible broken membrane at the end of the etching when the wafer is in the holder. It seems to be not critical since the solution cannot damage the circuits on the front side but the etchant could leave some unsightly white films on the front side which are not dissolved during the etching.

It must be noted that during the silicon dissolution, it is necessary to control that there is no overpressure in the bath caused by the exothermic reaction between the TMAH and the silicon. Finally, the boiling DI water is used to stop the chemical reaction and to dissolve the remaining products of the reaction. When the holder is used, the rinsing in boiling DI water helps to unstick the wafer from the seal without wafer breakage.

1.6 Selectivity versus other metals

The selectivity versus other metallic layers was also studied. Each metallic layer was deposited by evaporation on oxidized silicon wafers and patterned by lift-off. Thicknesses were chosen typically around 100 nm. Samples were then immersed during 1 hour in pure TMAH (20 % at 90°C) as well as in TMAH with additives (TMAH 5 % at 90°C with required amounts of Si powder and APS). Results are summarized in Table 1.6. We observed that the pH reduction (occurring with additives in the TMAH solution) did not have any significant effect on these metallic layers. Chromium and Titanium were used as adhesion layers for Platinum and Gold in thicknesses around 5 and 10 nm. Gold on Chromium was observed fully inert but Platinum on Chromium revealed some adhesion problems after TMAH etching due probably to TMAH diffusion between both layers. When Titanium replaced Chromium for this purpose, Titanium was observed underetched in pure TMAH as well as in TMAH with additives.

Table 1.6: Summary of metalic layers selectivity in pure TMAH and in TMAH with additives. √ means that the solution can be considered without effect on the metal; × if the solution damages the film.

Metal layer	pure TMAH	TMAH + additives
Gold	√	√
Chromium	√	√
Titanium	×	×
Platinum	√	√
Tungsten	×	×

Finally, our tests with Tungsten revealed some adhesion problems on oxide. Therefore, tests were achieved on silicon but would need to be confirmed on oxide with an optimized Tungsten deposition.

1.7 Etch-Stop

To micromachine structures with high accuracy, the only control of the etching time for a given couple concentration-temperature is hazardous due mainly to the nonuniformity of the silicon wafer thickness (from one to another wafer as well as from one position to an other on a same wafer). It is reported in [45] that wafer nonuniformity leads to large variations in membrane thickness (a 3" wafer 380 μm thick having nonuniformity of 2 μm on the entire surface leads to membrane thickness variations of more than 5 μm for 20 μm thick membrane). Furthermore, an etch-start delayed by native oxide on silicon can also falsify the time control. The total amount of etched silicon can also have an impact on the etch rate (called the loading effect) [27]. Finally, organic contaminations of TMAH can drastically decrease its etch rate. It is therefore critical to be able to stop silicon etching when a required cavity depth or a certain membrane thickness is reached [27].

1.7. ETCH-STOP

Three etch-stop techniques will be described here: dielectric etch-stop, p^{++} (silicon as polysilicon) etch-stop and electrochemical (or p-n junction) etch-stop. They are depicted in Fig. 1.10 in the special case of SOI technology.

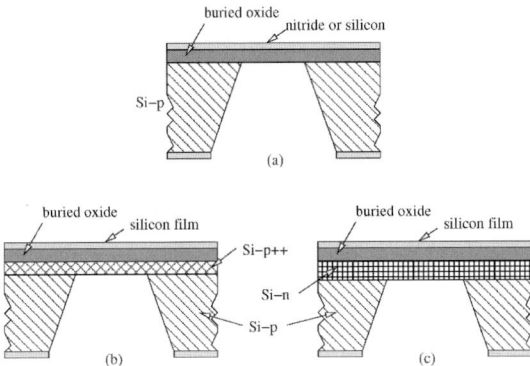

Figure 1.10: Etch-stop techniques; (a) dielectric etch-stop, (b) p++ etch-stop and (c) electrochemical etch-stop (the Si-n layer needs to be contacted in this case).

The simplest way is to stop the etching on a dielectric layer such as thermal oxide or LPCVD nitride which shows great selectivity versus silicon in TMAH. The buried oxide in a SOI wafer sandwiched between two layers of crystalline silicon, is an especially good way to achieve membranes with very accurate thicknesses (Fig 1.10(a)). In our case where we want to fabricate dielectric membranes, bulk silicon is etched to the buried oxide and since silicon film was previously removed on top and replaced by LPCVD nitride, we obtained very thin and controlled oxide-nitride sandwiches. But higher membrane thicknesses could also be obtained without removing the SOI top silicon film which is available in a wide variety of thicknesses. If membrane is still not thick enough, following methods can be advantageously combined with SOI technology.

As reported in [11] and [46], silicon etch rate in TMAH is reduced by 10 in regions doped with boron to a concentration of $\geqslant 10^{20} cm^{-3}$ (in a 22 % TMAH solution at 90°C) and by 40 to a concentration of $\geqslant 2 \times 10^{20} cm^{-3}$ in a 25 % TMAH solution at 80°C (note that $2.5 \times 10^{20} cm^{-3}$ is the solid solubility limit for boron in silicon at around 1200°C [11][46]). Heavy doped silicon can therefore be used as etch-stop for TMAH etching. To extend results reported in literature, measurements of the selectivity between p^{++} doped and undoped silicon were performed in our lab in various solutions at 80°C [47]. Boron implantations were performed at high energy (120 keV) and the doping profile through the silicon was confirmed constant by simulations after diffusion (5 hours at 1100°C) over a few tenths of microns before logarithmically decreasing. In order to get an accurate idea of the etch rate, the etching was therefore performed in the range in which the concentration was constant. A selectivity of 10 to 1 was achieved in a 25 % TMAH solution in silicon doped at $6 \times 10^{19} cm^{-3}$. In a 20 % TMAH solution, this concentration became $10^{20} cm^{-3}$. These results agree well with literature results reported above. Heavy doped polysilicon can also be used as etch-stop layer as demonstrated for the first time in [47]. The selectivity was shown the same for polysilicon and silicon allowing identical uses. But the advantage of polysilicon is that it can be deposited at any process step and does not require clever epitaxy steps as for silicon. A drawback of p^{++} silicon etch stop is that high boron concentrations in silicon can introduce tensile stress since boron is introduced substitutionally into the Si lattice, and boron atoms are smaller than silicon [27][11]. A solution is to add some germanium to compensate the tensile stress since it is a larger atom than silicon. Silicon doped with B and Ge still etches much slower than undoped silicon and stress in the layer is therefore reduced [27]. Another disadvantage of this method is that extremely high boron concentrations are not compatible for piezoresistive pressure sensors as well as for CMOS circuitry put on the membrane so they can only be used for microstructures without integrated electronics. This incompatibility can be avoided by growing an epitaxial layer of lighter doped Si on top of the highly doped boron layer [27] or by using SOI technology. In this case, a high boron doped region can indeed be implanted underneath the buried oxide to increase the membrane thickness without neglecting CMOS compatibility (Fig. 1.10(b)).

The last technique is commonly called electrochemical etch-stop. We will not fully detail it here as it is widely explained in [27][1][22][48][49]. In this case, a lightly doped p-n junction is used as an etch stop by applying a bias between the wafer and a counter electrode in the etchant [27]. The disadvantage of this technique is that the wafer must be mounted in an expensive mechanical holder to provide the aluminum contact. But its great advantage in comparison with the previous one is its full CMOS compatibility. Furthermore, it allows thicker doped profiles due to the lower concentration and therefore higher membrane thicknesses. This technique can also be used in SOI technology to increase the membrane thickness as demonstrated in [50] to perform high temperature pressure sensors. In this case again, a n-type layer is implanted underneath the buried oxide in a depth depending of the required thickness (Fig. 1.10(c)). Unfortunately, in case of SOI wafers, it is critical to perform the ohmic contact to the n-type layer through the buried oxide. But solutions are proposed by [51] and [52] which do not required the use

1.8. UNDERCUTTING

of an externally applied bias voltage. The process in this case is based on the formation of a wet battery when a gold-chrome/n-type silicon/TMAH construction is formed.

1.8 Undercutting

1.8.1 Generalities

As previously introduced, a cavity will faithfully fit the patterned opening if the opening is rectangular or square and has only concave corners and sides perfectly aligned with ⟨110⟩ directions. In other cases, undercutting will inevitably occur.

Undercutting occurs especially when a square mask having only **concave** (or internal) corners is misaligned with the [110] direction (Fig. 1.2(b) and Fig. 1.11(a)). This misalignment results in an oversized opening, oriented in the [110] direction and bounded by four {111} planes that encompass the original design [45][13]. More generally, a mask opening with arbitrary closed geometry and orientation (for example a circle) after sufficiently long etch time will result in a rectangular pit in the silicon circumscribing the mask opening, bounded therefore by the {111} surfaces and oriented in the [110] direction (Fig. 1.3(a)) [27].

(a)   (b)

Figure 1.11: Undercutting in case of misalignment (a) and at convex corners (b).

On the other hand, undercutting occurs irrespective of alignment conditions in presence of **convex** corners, i.e. corners turning outside or $> 180°$ [27]. Convex corners in a mask opening will always be completely undercut by TMAH after sufficiently long etch time as shown on Fig 1.3(b) and Fig. 1.11(b). This undercutting can be explained by the fact that fast etching planes ({314}planes in case of TMAH [29]), in addition to the

{100} planes, become easily exposed at convex corners, allowing faceted etching at those locations [13].

Undercutting due to misalignment of concave corners can be disadvantageous to etch membranes with accurate widths but can be advantageous for undercutting suspended bridges (Fig. 1.12(a)). On the same way, undercutting at convex corners can be a benefit for releasing suspended cantilevers (Fig. 1.12(b)) but is a problem when attempting to create a mesa rather than a pit.

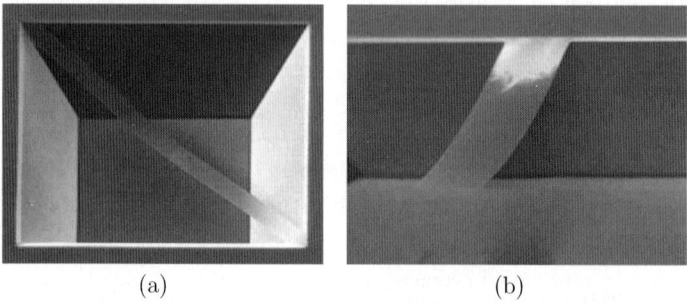

(a) (b)

Figure 1.12: Suspended bridge (a) and cantilever (b) taking into account the advantages of the undercutting.

If mesa is required, undercutting at convex corners can be reduced or even prevented by corner compensation structures which are added to the corners in the mask layout [27]. Rectangular or square patches, triangular patterns or rotated rectangles can be placed at the convex corners of the mesa to slow the exposure of fast-etching surfaces as detailed in [27]. This technique needs however to be carefully designed by simulations [53] and an accurate control on the etching time. We will not detail it here.

An interesting demonstrator combining, in one structure, misalignments and undercutting at convex corners was performed. This structure is used in some gas sensors applications. The thermal oxide on the silicon is patterned as a central square supported by four tethers in diagonal fashion [1]. When TMAH etching begins, undercutting of the four tethers takes place because they are not aligned to the $\langle 110 \rangle$ directions. The tethers are fully undercut, creating four convex corners. This in turn causes further undercutting of the oxide that is patterned as a central square. Fig. 1.13(a) shows the structure after 2 hours of etching; with silicon still supporting the oxide in the center. When the etching is finished after 3 hours (Fig. 1.13(b)), the central square is fully undercut and suspended. Buckling of the tethers is caused by compressive stress in the oxide [1].

1.8. UNDERCUTTING

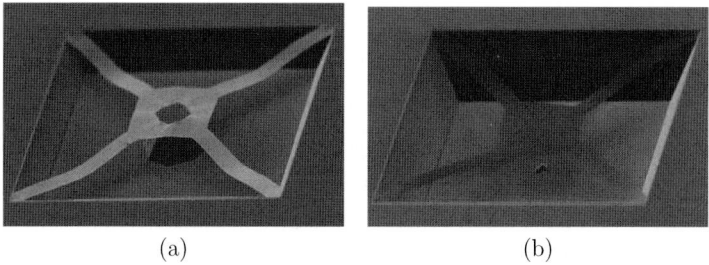

Figure 1.13: Undercutting of a suspended 400 nm thick membrane structure supported by four tethers; (a) after 2 hours and (b) after 3 hours in TMAH 25% at 95°C.

1.8.2 Membranes patterning

In case of dielectric membranes released by back side etching, undercutting induced by misalignment will affect the actual membrane dimensions. Therefore, it must be approximately evaluated to be taken into account.

Etching accurately a square diaphragm needs to align it at best to the flat of the wafer since this flat indicates the [110] direction. Nevertheless, the flat itself can be misaligned to 1° (0.5° in the best cases) with respect to the [110] direction as specified by the manufacturers [27]. If the alignment of a pattern with the [110] direction is really critical, pre-etch alignment targets can be used to delineate the planes of interest as proposed in [27]. Alignment with better than 0.05° accuracy could be accomplished this way [27]. This undercutting can also be due to the finite etching of the {111} planes since for TMAH the selectivity between {100} and {111} surfaces was reported to be 35 to one in best cases (TMAH 5 % at 90°C [21]). As showed in [21] and [36], this selectivity decreases with increasing the TMAH concentration and decreasing temperature (selectivity of 10 for one was observed at TMAH 40 % and 70°C [21]). But as demonstrated in [27], the knowledge of the exact etch-rate in the [111] direction is not so usefull since in practice, the undercutting due to this etch-rate is much smaller than the undercutting due to the misalignment. The total undercutting will therefore be mainly determined by misalignment, rather than by etching of {111} walls [27].

When membranes are built by back side etching in post-process as in our case, a double side mask alignment must be performed. The first mask of the IC process (membrane location definition in case of a gas sensor process in SOI technology) needs to be accurately aligned with the flat of the wafer. The following masks must then be correctly aligned on this first mask. At the end of the process, the back side mask is finally aligned with as much precision as possible, with the last front mask (currently metal layer). The total misalignment in this case is the sum of misalignments occurring during each photolithography step.

Therefore, the membrane width a will be completely defined by the thickness of the wafer t (200±25 μm in most cases), the width of the back side mask opening b, the slope of the anisotropic etching (54.74°) and the undercutting due to the etching in the [111] direction and to the total misalignments of the back side mask opening versus the [110] direction. The undercutting can be taken into account such as an average empirical correction factor λ. The size a of the membrane can therefore be written as

$$a = b - 2\,cotang\,(54.74°)\,t + \lambda = b - e\sqrt{2} + \lambda \qquad (1.7)$$

The average empirical correction factor λ including the undercutting effects needed to be evaluated by experiments. We evaluated it through 5 membranes per wafer on more than ten 200 μm thick micromachined wafers in several etching solutions. Underetch values from 7.4 to 10.5 μm on each side of the cavities were extracted with more measured values around 9.3 μm. Therefore, the empirical correction factor was fixed equal to 9 μm. So, to build for instance a 400x400 μm^2 membrane on a 200 μm-thick wafer, the back side opening must be designed equal to 674x674 μm^2. The backside opening takes therefore 137 μm more on each side in comparison with the membrane size. This value reveals that bulk micromachining involves extensive die area consumption. It can result in chips quite fragile if not enough silicon remains unetched around the cavity. To improve compactness, dry etching can be interesting. Thank to its high aspect ratio, this technique enables to integrate more membranes per unit of area.

Without correction factor, the membrane would be about 10 μm wider than required. But even with this correction, the reproducibility from one wafer to another can not be fully guaranteed due to the lack of precision on the wafer thickness and the probable difference of alignment between each wafer. Overetch can also lead to additional undercutting. Furthermore, it is difficult to provide membranes perfectly located regarding a given pattern on top side due to the relative precision of the backside alignment. It can be critical especially in case of pressure sensors where transistors are used as sensing part. Dry etching can again constitute a more appropriate technique in this case thanks to its high aspect ratio. But in our case where 200 μm-width microheaters need to be centered on 400x400 μm^2 or 600x600 μm^2, the alignment accuracy as well as the patterning is widely sufficient.

1.9 Summary

From various anisotropic silicon etchants, we focused on TMAH, the better etchant fully adapted to our applications. Its CMOS compatibility, its high selectivity versus dielectrics as well as aluminum gave this etchant very attractive to release membranes in post-processing without special care regarding the integrated circuits.

Its properties of silicon etch rate and roughness were studied in details. Its selectivity versus the dielectrics in contact with the etchant solution during the membrane release was quantified. Therefore we determined the best solution allowing fast and smooth silicon etching through 200 μm without damaging aluminum. It is based on a TMAH 10 % solution heated at 90°C and doped with 35 g/l of silicon powder and 15 g/l of APS.

How to accurately stop the silicon etching was then detailed. In case of our dielectric membranes built in SOI wafers, the etching is easily stopped on the buried oxide. If thicker silicon membranes are required, we explained that other solutions as well as heavily boron doped silicon or p-n junctions are available and can be combined with CMOS-SOI technology.

We then focused on the undercutting property. We showed how to advantageously use it to process suspended microstructures as bridges or cantilevers. These microstructures will be useful to quantify the residual stress included in most of dielectrics films. Finally, we demonstrated that undercutting can prevent to obtain accurate membranes dimensions. Therefore we quantified this undercutting to take it into account in our membranes design.

Chapter 2

Thin dielectric films stress extraction

Residual stress in thin films is a major concern for the operation and the reliability of MEMS. Various methods have been published these last years to extract the residual stress in polysilicon [27][54][55][56][57]. Nowadays, other types of materials such as SiO_2 or Si_xN_y are also being used in MEMS as structural material, especially in the fabrication of suspended membranes for pressure or microheaters based sensors. Residual stress can be compressive, which makes the film expand parallel to the surface, or tensile, which makes the film shrink. The profile as well as the robustness of a thin dielectric membrane are affected by the residual stress. A too compressive membrane may buckle, avoiding its use to support sensors. A too tensile membrane can break in presence of pressure or high temperature gradients. It is therefore essential to be able to measure and control the stress separately in each layer composing the structure in order to design reliable multilayers.

The use of stress values extracted from the literature is not sufficient since materials properties can change a lot from one piece of equipment to another as well as between two processes. It is therefore not only necessary to characterize the silicon oxide and the silicon nitride produced in each laboratory but also essential to be able to measure the stress during each fabrication batch on processed wafers.

This chapter reports our detailed study on the residual stress in the thin dielectric films constituting a typical released membrane: thermal oxide, LPCVD nitride and PECVD oxide[1]. Residual stresses in oxide have not been much studied so far since oxide is mainly used as a sacrificial layer for the release of structures in polysilicon, nitride or metal or as a non dominant part of a thick silicon membrane. Stress measurements in nitride films, on

---
[1]Oxide and nitride will be used from now on as abbreviated forms of respectively silicon oxide and silicon nitride.

the other hand, are more common and a lot of methodologies are available. Nevertheless, measurements of these dielectric films at very low thicknesses as in their actual use are not so common. And stress measurements for the combinations of these layers are even more rare.

Two measurement methods are presented and compared in this chapter. The first one is the well known substrate curvature technique [27], generalized to extract the residual stress in thin stacked films. This technique also demonstrates how to extract the residual stress in the buried oxide of a SOI wafer and in the membrane at the end of our complete gas sensor process. The second one is based on micromachined microstructures [54] designed for our purpose, i.e. characterization of the stress in very thin dielectric layers. The anisotropic silicon etching properties of the CMOS compatible etchant TMAH were used to release strain measurement structures made in thin films of oxide, nitride, as well as a combination of each of them on silicon. The variation of stress throughout the membrane thickness was finally investigated with dedicated microstructures.

But in order to support the correct interpretations of our experimental results, we think interesting to firstly summarize the main theoretical concepts of thin films mechanics useful for MEMS researchers and designers.

2.1 Introduction - Definitions

2.1.1 Stress, strain and Elastic constants

Stress σ is the force per unit area that is acting on a surface of a solid, more commonly expressed in Pascals (Pa) or N/m^2.

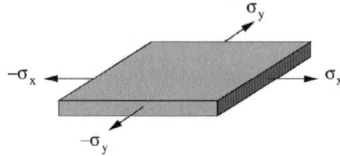

Figure 2.1: Stress on a differential volume in static equilibrium.

In case of a thin film on a thick substrate, no stress will occur in the direction z, normal to the substrate, as a film under stress can only expand or contract by bending

2.1. INTRODUCTION - DEFINITIONS

the substrate and deforming it in a vertical direction [27]. The plate is thin enough to avoid any variation of displacement with respect to z. Vertical deformations will not induce stresses in a substrate because it freely moves in that direction. We call this state plane stress. In such a biaxial system (x, y), the shear stress (acting along the surface) can also be eliminated [58]. We finally assume that no significant external forces act on the system. This means that the surface forces on opposite sides of the structure must be equal as shown in Fig. 2.1. In other words, the system is in static equilibrium. The two elements of stress can be therefore represented as the following matrix

$$\sigma = \begin{pmatrix} \sigma_x & 0 \\ 0 & \sigma_y \end{pmatrix} \tag{2.1}$$

Stresses can be the result of externally applied forces. In this case, after the load is removed, the stresses are expected to vanish. On the other hand, thin films are stressed even without the application of externally imposed forces and are characterized by an internal or **residual stress** [59] as we will explain further. In the following sections, all stresses will be supposed to be residual stresses. Residual stress can be compressive or tensile. By convention, stress in first case will be expressed with a minus sign and in second case, with a positive sign.

When subjected to a stress, any free body literally get pushed (or pulled) out of shape. **Strain** ε is a measure of this deformation [11]. It is therefore defined in terms of the partial derivatives of the displacement. Strain is a dimensionless variable. In static equilibrium, the strain is caused by stress, so the principal coordinate system is the same for stress and strain [58].

Materials of interest (silicon, oxide and nitride) being completely elastic (and no plastic), they obey to Hook's law; that means, they deform linearly with load. Since load is proportional to stress and deformation is proportional to strain, stress and strain are linearly related:

$$\begin{pmatrix} \varepsilon_x \\ \varepsilon_y \end{pmatrix} = \begin{pmatrix} H_{11} & H_{12} \\ H_{21} & H_{22} \end{pmatrix} \begin{pmatrix} \sigma_x \\ \sigma_y \end{pmatrix} = H \begin{pmatrix} \sigma_x \\ \sigma_y \end{pmatrix} \tag{2.2}$$

From our previous assumptions, H matrix has only two independent elements or two elastic constants: the elastic modulus or Young's Modulus E and the Poisson ratio ν.

$$H = \begin{pmatrix} 1/E & -\nu/E \\ -\nu/E & 1/E \end{pmatrix} \qquad (2.3)$$

Eq.2.2 can therefore be reduced to the following strain-stress relationships :

$$\varepsilon_x = \frac{1}{E}(\sigma_x - \nu\,\sigma_y) \qquad (2.4)$$
$$\varepsilon_y = \frac{1}{E}(\sigma_y - \nu\,\sigma_x)$$

In an isotropic material[2], $\varepsilon = \varepsilon_x = \varepsilon_y$ so that the in-plane or **biaxial stress** $\sigma = \sigma_x = \sigma_y$ is equal to

$$\sigma = \left(\frac{E}{1-\nu}\right)\varepsilon \qquad (2.5)$$

The ratio $E/1-\nu$ is called the **biaxial modulus**.

This equation is valid for the specific boundary conditions seen by a thin film on a rigid substrate. If the film was subjected only to **uniaxial stress** (i.e. the film is restricted to move only in one direction), the stress is simply

$$\sigma = E\,\varepsilon \qquad (2.6)$$

However, this equation can rarely be used in presence of internal residual stress since stress in this case is a thermal or growing effect in all directions instead of a traction or contraction due to an external load.

The **elastic modulus** or **Young's Modulus** E tells us how much a material is elongated under a given load. It has units of force per unit of area, just like stress and pressure. It may also be thought as the stiffness or material resistance to elastic deformation. The higher the material elastic modulus, the lesser it deforms for a given stress, and thus the stiffer it is. For example, an incompressible material would have an infinite Young's modulus, while a "soft" material would deform considerably for a given amount of stress, so its modulus of elasticity would be quite low [11].

Under uniaxial tensile stress, the element expands in the direction of the stress, and contracts in directions perpendicular to the stress. In this situation, there are two strains,

---

[2]Or in a cubic crystal like silicon in case of our previous assumptions.

2.1. INTRODUCTION - DEFINITIONS

one axial and one transverse. The ratio of the transverse to the axial strains, or the proportionality between the contraction and the elongation, is the **Poisson ratio** ν. This ratio is dimensionless, and has a value between 0.2 and 0.5 for most materials. The volume of the element has changed as a consequence of the strain. The volume expansion is proportional to $(1 - 2\nu)$, which means that materials with $\nu = 0.5$ does not change their volume under uniaxial stress and are called incompressible [58][11].

Table 2.1 gives the elastic constants generally accepted for some common MEMS materials. The Young's modulus and Poisson ratio of silicon are dependent on its crystallographic orientation and can vary according to the direction in a given plane (such as in {100} planes). However, biaxial modulus is constant in the (100) plane as demonstrated in details in [60] and is equal to **180.5 GPa**. The silicon Young's modulus and Poisson ration in the ⟨100⟩ crystallographic directions (as cited in Table 2.1) are commonly measured by various methods and calculated in [61][33] or mentionned in [13][62]. These values are more consistent than a lot of others cited without reference to any crystallographic plane (as in [58][27][11]). Furthermore, the silicon Young's modulus values calculated by nanoindentation (as in [63]) are not convincing since they depend on the silicon doping level and the applied load. That was also confirmed by measurements performed in our lab. As explained in [64], the silicon phase changes during indentation and results in a modification of the Young's modulus value.

Table 2.1: Mechanical properties of some important MEMS materials.

Material	Young's modulus [GPa]	Poisson ratio [s.d.]	Source
Silicon (100)	130	0.28	[60][61][33][13][62]
Polysilicon	160	0.23	[65][27][59][58][11]
Silicon dioxide	70	0.20	[65][27][59][58][11]
Stoichiometric LPCVD nitride	270	0.27	[65][27][59][58][11]
Aluminum	70	0.35	[65][27][59][58][11]

Elastic constants of the other materials mentioned in Table 2.1 are less object of discussion. Nevertheless, the commonly available tabulated values of mechanical properties of materials are generally derived from bulk specimens, and thus may not be very relevant to the materials and scales used in micromachined devices [11]. But good approximations can be made with these available data.

Existing methods to extract elastic modulus and Poisson ratio in thin films made of harder materials are the bending beam and resonant beam methods [62][1]. The first one consists in loading the tip of a cantilever beam using a nanoindenter (as is called a load controlled submicrometer indentation instrument [1]) and to measure the force-displacement curve to finally extract the Young's modulus [62]. In the second method, the Young's modulus is obtained by measuring the resonant frequency of a beam under excitation [62]. Both methods can also advantageously be used to extract the Young's modulus for each layer of multi-layered cantilevers without accessing the individual layers during processing as demonstrated in [66] and [65]. The Poisson ratio for thin films is more difficult to measure than the Young's modulus as thin films tend to bend out of plane in response to in-plane shear [27]. Fortunately, it is not required to know Poisson ratio in order to extract stress in most microstructures as we will see further.

As process parameters greatly influence these elastic properties and may be difficult to reproduce on one set of equipment, let alone on the equipment of others [11], they might be usefully measured in the thin films under consideration. Nevertheless, the Young's moduli of the layers under consideration in our work are generally well known and will be used here.

2.1.2 Uniform and non-uniform stresses and strains in thin films

Residual stress can be uniform or non-uniform through the depth of a thin film. If the stress is uniform, its measurement will give an **average stress**. If the stress is non uniform, a difference of stress or **stress gradient** exists between the top and the bottom of the thin film. Using dedicated methods, this stress gradient can be isolated and measured, otherwise, the measured stress will be an average stress through the depth taking into account the vertical variation of stress.

Uniform stress will produce a deformation and therefore a proportional **average relaxed strain**. In the same way, a stress gradient will lead to a **relaxed strain gradient**. If a body is stressed but not free to move, it will not be allowed to relax, and no deformation and therefore no relaxed strain will be able to occur. The body is said under residual stress and therefore under residual strain proportional to the residual stress, but without relaxed strain. If the same stressed body is now allowed to move, it will lead to relax until an equilibrium state, and a deformation will occur. In this case, the stress is vanished and was replaced by a relaxed strain. So, it is important to note our distinction between the **strain** equal to the stress on Young's modulus ratio and present in same time that the

2.1. INTRODUCTION - DEFINITIONS

stress, and the **relaxed strain** equal to the strain in absolute value but compensating the strain and therefore of opposite sign. Both strains are mathematicaly identical but the use of two different terms will facilitate its understanding. We will see along this chapter one technique to measure stress by transformation of stress into relaxed strain through the release of a microstructure.

Nearly all films have a state of residual stress, due to mismatch in the thermal expansion coefficient between the film and the substrate, to lattice mismatch, to substitutional or interstitial impurities, to growth processes, etc. [27]. All factors of stresses can be classified in two groups: **extrinsic** and **intrinsic** stresses.

Extrinsic stress

The extrinsic stress is imposed by unintended external factors such as temperature gradients. It is commonly uniform through the depth. **Thermal-mismatch stress** is the more common source of extrinsic stress and is well understood. It specially arises in structures with inhomogeneous thermal expansion coefficients, subjected to uniform temperature change.

To understand the causes and effects of thermal-mismatch stress in thin films, consider the typical structure shown in Fig. 2.2.

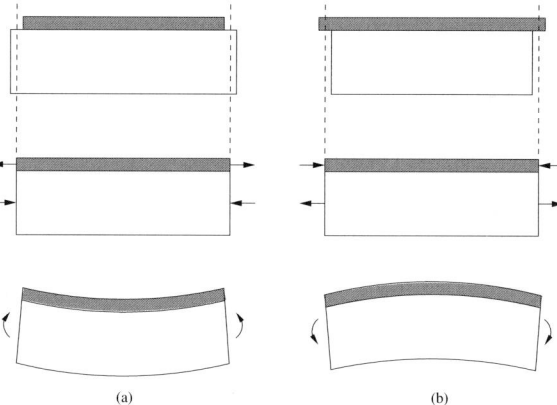

Figure 2.2: Sequence of events leading to (a) residual tensile stress in film; (b) residual compressive stress in film (adapted from [59]).

Considering, as it is often the case, that the thin film was deposited on the substrate at an elevated temperature. In general the thin film and substrate have different **thermal expansion coefficients**. The thermal expansion coefficient of a material is defined as

$$\alpha = \frac{d\varepsilon}{dT} \tag{2.7}$$

Assuming that the thermal expansion coefficients are temperature independent, the strain caused by thermal expansion is then simplify

$$\varepsilon(T) = \varepsilon(T_0) + \alpha \cdot \Delta T \tag{2.8}$$

where the first part of the equation is assumed to be negligible and the last part, the strain caused by thermal expansion.

When a thin film is deposited on a thick substrate at elevated temperature, and subsequently cooled and operated at ambient temperature, the difference between the thermal expansion coefficients of the film and substrate creates stress and strain.

First consider the behavior shown in Fig. 2.2(a) where the growing film initially shrinks relative to the substrate. Usually, the substrate dimensions undergo minor shrinkage in the plane while the film dimensions may reduce significantly. Compatibility, however, requires that both the film and substrate have the same length. Therefore, the film is constrained and stretches, while the substrate accordingly contracts. The tensile forces developed in the film are balanced by the compressive forces in the substrate. However, the combination is still not in mechanical equilibrium because of the uncompensated end moments. To remind you, equilibrium requires that forces (F) and bending moments (M) vanish on any film/substrate cross section. If the structure is free to move, it will elastically bend to counteract the unbalanced moments. Thus, films containing internal **tensile stresses** bend the substrate concave upward. Similarly, **compressive stresses** lead their origin from films that tend to initially expand relative to the substrate (Fig. 2.2(b)). Internal compressive stresses, therefore, bend the substrate convex outward [59].

As the substrate is very thick compared to the film, it is a good approximation to assume that the substrate contracts to the size it would have attained in absence of the film [58]. With this assumption, we can write the strain of the substrate as

$$\varepsilon_s = -\alpha_s \cdot \Delta T \tag{2.9}$$

where α_S represents the coefficient of thermal expansion for the substrate and the minus

2.1. INTRODUCTION - DEFINITIONS

sign, the compression of the film. The film then gets this same strain due to the fact that it is attached to the substrate.

$$\varepsilon_{f,attached} = -\alpha_s \cdot \Delta T \qquad (2.10)$$

If the film was unattached, however, its strain would be

$$\varepsilon_{f,free} = -\alpha_f \cdot \Delta T \qquad (2.11)$$

where α_f represents the coefficient of thermal expansion for the thin film.

The difference between the strains film features with and without attachment to the substrate is the thermal mismatch strain [58]

$$\varepsilon_{f,mismatch} = \varepsilon_{f,attached} - \varepsilon_{f,free} = (\alpha_f - \alpha_s) \cdot \Delta T \qquad (2.12)$$

The thermal mismatch leads to stress in the film. From previous equation, we can write in case of biaxial system

$$\sigma_{f,mismatch} = \frac{E}{1-\nu}(\alpha_f - \alpha_s) \cdot \Delta T \qquad (2.13)$$

By convention, tensile stress is positive, compressive stress is negative. Therefore, if $\alpha_f < \alpha_s$, a compressive stress will be expected and if $\alpha_f > \alpha_s$, a tensile stress will appear.

Intrinsic stress

The intrinsic stress reflects the internal structure of a material during its deposition. It is less clearly understood than the thermal stress. It depends on deposition rate, deposition temperature, pressure in the deposition chamber, incorporation of impurities during growth, grain structure, fabrication process defects, etc. In most cases, intrinsic stress is non uniform through the depth and is therefore responsible of stress gradient.

For instance, the phosphorous doped polysilicon is expected to be more compressive than pure polysilicon because the phosphorous atom is larger than silicon [27]. Boron is another dopant in silicon which exerts a tensile stress when introduced into the crystal lattice. As the smaller boron atom displaces the silicon atom, there is a tendency for the lattice to contract locally. However the silicon lattice will restrain from contracting and therefore results in a local tensile stress. High level of boron concentration is usually used as etch stop for controlling the depth in silicon substrate of the height of a silicon beam which generates high levels of stress gradient since the vertical dopant profile is not uniform and causes therefore a stress distribution varying with depth [67]. During layers deposition, microvoids can also appear when by-products escape as gases and the lateral diffusion of atoms evolves too slowly to fill all the gaps, resulting in a tensile film [27]. Finally, thermal oxide shows stress gradient due to oxygen diffusion while the film is forming [68] but is in same time under compressive stress due to the fact that in the oxide layer one silicon atom takes nearly twice as much space as in single crystalline silicon [27].

Intrinsic stress can sometimes be annealed out completely but the anneal temperatures are quite high and may not be practical for the production of micromechanical devices [13]. On the contrary, some amount of thermal mismatch stress is unavoidable when working with materials having different thermal expansion coefficients.

2.2 Stress measurements by substrate curvature method

2.2.1 Theory

We consider a composite plate film/substrate of width w [59][69] as in Fig. 2.3.

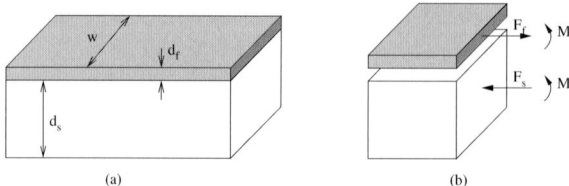

Figure 2.3: Stress analysis of film/substrate combination: (a) structure; (b) free-body diagrams of film and substrate with forces and end moments (adapted from [59]).

The film thickness and Young's modulus are d_f and E_f, respectively, while the corresponding substrate values are d_s and E_s. Due to internal stresses, mismatch forces arise at the film/substrate interface. As seen on Fig. 2.3, each set of forces can be replaced by the static equivalent combination of a force and moment; F_f and M_f in the film, F_s and M_s in the substrate, where $F_f = F_s$ since film and substrate are sticked together. Force F_f can be considered to act uniformly over the cross-sectional area $d_f.w$, or on the middle

2.2. STRESS MEASUREMENTS BY SUBSTRATE CURVATURE METHOD 57

of this section at $\frac{d_f}{2}$. In this case, equilibrium requires that

$$F_f . \frac{d_f}{2} = M_f \qquad (2.14)$$

or for the complete structure

$$F_f . \frac{(d_f + d_s)}{2} = M_f + M_s \qquad (2.15)$$

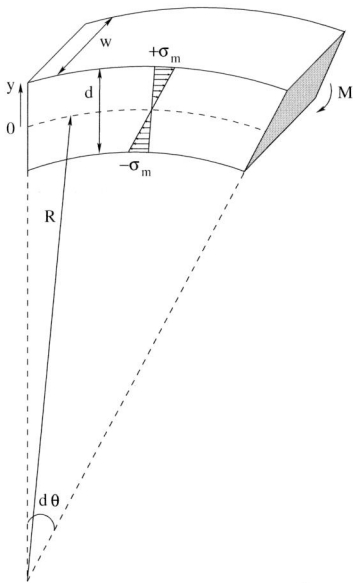

Figure 2.4: Elastic bending of a beam under applied moment (adapted from [59]).

Considering now an isolated beam segment bent by a moment M (Fig. 2.4), the deformation is assumed to entirely consist of the extension or contraction of longitudinal beam fibers by an amount proportional to their distance from the central or neutral axis, which remains unstrained in the process. The stress distribution reflects this by varying linearly across the section from maximum tension $+\sigma_m$ on the top to maximum compression $-\sigma_m$ on the bottom [59].

The length of the neutral axis where the stress equals zero, is given by

$$L_{y=0} = R.d\theta \tag{2.16}$$

where R is the radius of curvature of the beam segment and $d\theta$, the subtended angle. The length of the section at an arbitrary position y is

$$L_y = (R+y).d\theta \tag{2.17}$$

The strain along the y axis is the difference between the length of the neutral axis and the length at position y

$$\varepsilon_y = \frac{L_y - L_{y=0}}{L_{y=0}} = \frac{(R+y).d\theta - R.d\theta}{R.d\theta} = \frac{y}{R} \tag{2.18}$$

The axial stress is then given by combination of 2.6 and 2.18

$$\sigma_y = E\frac{y}{R} \tag{2.19}$$

and the maximum stresses on the top and bottom of the beam are therefore equal to

$$\sigma_m = \pm E\frac{d}{2R} \tag{2.20}$$

showing that compressive stress on the top is negative and tensile stress on the bottom is positive.

We can now calculate the bending moment of the beam segment by integrating the stress over the beam section.

$$M = 2\int_0^{d/2} \sigma_y \, w.y \, dy = E\frac{d^3 w}{12R} = \frac{E}{R}I \tag{2.21}$$

where $I = \frac{wd^3}{12}$ is called the **moment of inertia** of a rectangular beam with respect to the center of the beam in a direction perpendicular to one of the sides [69]. By extension of this result, we have

$$M_f = E_f\frac{d_f^3 w}{12R} \quad \text{and} \quad M_s = E_s\frac{d_s^3 w}{12R} \tag{2.22}$$

Finally, in order to account for actual biaxial-stress distribution in films, rather than the uniaxial stresses assumed for ease of integration, it is necessary to replace E_f by $E_f/1-\nu_f$

2.2. STRESS MEASUREMENTS BY SUBSTRATE CURVATURE METHOD

and similarly for E_s. Substitution of these terms in Eq. 2.15 yields

$$F_f \cdot \frac{(d_f + d_s)}{2} = E_f \frac{d_f^3 w}{12R(1-\nu_f)} + E_s \frac{d_s^3 w}{12R(1-\nu_s)} \tag{2.23}$$

Since d_s is normally much larger than d_f, the equation becomes

$$F_f \cdot \frac{d_s}{2} = E_s \frac{d_s^3 w}{12R(1-\nu_s)} \tag{2.24}$$

Furthermore, since σ_f is equal to the force F_f acting on the area wd_f or $\sigma_f = F_f/wd_f$, the film stress σ_f can be then given by

$$\sigma_f = \frac{E_s d_s^2}{6(1-\nu_s)} \cdot \frac{1}{d_f} \cdot \frac{1}{R} \tag{2.25}$$

This is the **Stoney equation**. Values of stress in the film are determined through measurement of curvature R if other parameters are known.

Stress measured by Stoney method is an average stress since it is obtained by integrating the stress over the section or more precisely, over the thickness of the beam. If the thin film has an intrinsic stress or stress gradient, the Stoney calculated stress will be the average stress present at the middle of the beam, at the neutral axis, taking therefore into account, the effect of the gradient.

Compressive and tensile stresses in curved substrate vanish since stress is relaxed when bending. The stress is replaced by relaxed strain expressed in this case by the curvature of the substrate. Thanks to this curvature, the stress can be measured. Without it, it would be impossible to extract the residual stress because the film would stay constrained. The curvature measurement therefore gives the residual stress the thin film would have if it was not able to be relaxed.

Thanks to the simplifications ($d_s \gg d_f$), only silicon Young's modulus and Poisson ratio must be known to extract the residual stress in any kind of thin films whatever the elastic properties of the film (E, ν and α values). It makes this method very useful to measure residual stress in a thin film when we do not know their elastic properties. On the contrary, if strain in the thin film must be extracted, its Young's modulus and Poisson ratio must be known.

Without the previous simplification, Stoney could be more rigorously written as

$$\sigma_f = \frac{1}{6R(d_f + d_s)d_f} \left[\frac{E_f d_f^3}{(1-\nu_f)} + \frac{E_s d_s^3}{(1-\nu_s)} \right] \quad (2.26)$$

requiring to know the complete elastic properties of the thin film we want to characterize. Calculations with experimental results in next part will demonstrate the validity of the previous simplification.

Validity of the Stoney equation implies that the substrate has transversal isotropic elastic properties with respect to the film. Using single crystal silicon substrates featuring moderately anisotropic properties ($\langle 100 \rangle$ oriented wafers) satisfies this transverse isotropy argument [27]. Films must also be uniform in thickness and stress must be homogeneous and equi-biaxial over the entire substrate. The legitimacy of these assumptions depends on the deposition process. Films such as thermal oxide, LPCVD nitride or PECVD oxide under consideration have a great uniformity unlike sputtered films where homogeneity is precarious [27].

About the thin film approximation, we can read in [70] that Stoney equation is a good approximation for thicknesses ratio d_f/d_s smaller than 10%. This equation seems to fail for properly describing the variation of stress with thickness and cannot be used for thickness ratios larger than 10%. The cited paper proposes modifications to the original Stoney formula which does not require information on the layers modulus, to improve calculations for thickness ratios up to 40%. As we will see further, thickness ratio in our case will never exceed 0.3% and confirm our ability to use the original Stoney equation.

For a stack of n layered and continuous films of thickness d_{fi} on a substrate with a thickness d_s always greater than the total films thickness $\sum_1^n d_{fi}$, the total curvature is just the sum of the individual films contributions since moments are additive [59]. It therefore appears that each film independently interacts with the substrate without accounting for the presence of adjacent films or the stacking sequence of films in the composite structure [71]. The stress σ_{fi} in the film number i therefore yields

$$\sigma_{fi} = \frac{E_s d_s^2}{6(1-\nu_s)} \cdot \frac{1}{d_{fi}} \cdot \frac{1}{R_i} \quad (2.27)$$

and the total stress σ_{total} in the n stacked films is

$$\sigma_{total} = \frac{E_s d_s^2}{6(1-\nu_s)} \cdot \frac{1}{\sum_{i=1}^n d_{fi}} \cdot \sum_{i=1}^n \frac{1}{R_i} \quad (2.28)$$

2.2. STRESS MEASUREMENTS BY SUBSTRATE CURVATURE METHOD

The expression shows that the total substrate curvature consists of a linear superposition of the bending effects resulting from each of the individual films. In other words, the stress in each film is proportional to the partial curvature in the substrate due to this particular film [71].

Our experimental results will confirm that this **superposition principle of Stoney** can be very useful to predict the internal stress value in stacked thin films on substrate when we know only the internal stress in each individual film and without knowing the elastic properties of the composite films.

2.2.2 Experimental results

The starting substrates were three inch diameter P-type $\langle 100 \rangle$ 380 μm thick bulk silicon wafers. Three different layers were grown on five silicon wafers as summarized in Table 2.2. The growth time of each identical layer was kept constant to allow accurate comparisons. The 400 nm thick thermal oxide was grown at 1000°C under a mixed O_2-H_2 atmosphere while the 300 nm thick LPCVD nitride was deposited at 800°C with a stoichiometric SiH_2Cl_2/NH_3 ratio of 0.33 to obtain a Si_3N_4 nitride. The 300 nm thick PECVD oxide was deposited in a plasma of $SiH_4+N_2+N_2O$ at 300°C and densified under O_2 at 800°C during 30 min. Since thermal oxide and LPCVD nitride growth in the furnace on both sides of our wafers, we etched the back side nitride layer by plasma and the back oxide (the top side being protected by a photoresist) using an HF solution. The thickness of the deposited films was measured using an ellipsometer.

Table 2.2: Three different layers deposited on five wafers.

Wafer	Layer
#1	Thermal oxide
#2	LPCVD nitride
#3	PECVD oxide (densified and not densified)
#4	Thermal oxide + LPCVD nitride
#5	Thermal oxide + LPCVD nitride + densified PECVD oxide

The residual average stress in each film was extracted comparing the wafer curvature before and after deposition using a Dektak profilometer over a distance of 50 mm with a stylus force of 20 mg. The profilometer gives the maximal deflection $h_{\max}$ of the wafer surface over the length of the scale [72] as shown in Fig. 2.5.

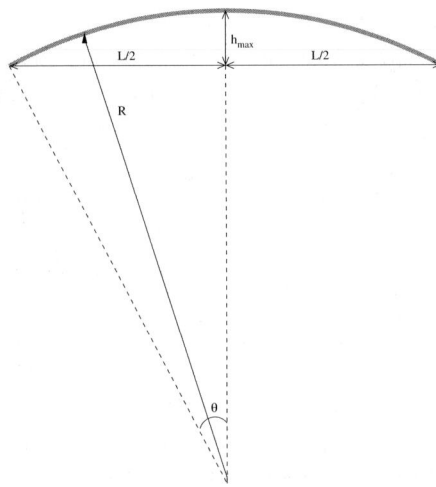

Figure 2.5: Evaluation of the radius curvature by profilometer scan.

From the schematic of Fig. 2.5, maximum deflection $h_{\max}$ can be expressed as a function of R

$$h_{\max} = R - R\cos\left(\frac{L/2}{R}\right) \tag{2.29}$$

For a small θ, $R \gg L$. With this approximation, $\cos\left(\frac{L/2}{R}\right)$ can be simplified in the two first factors of its Taylor series. Eq. 2.29 becomes

$$h_{\max} = R - R\left[1 - \frac{1}{2}\frac{(L/2)^2}{R^2}\right] \tag{2.30}$$

and the curvature radius R can therefore be expressed as

$$R = \frac{\left(\frac{L}{2}\right)^2}{2h_{\max}} \tag{2.31}$$

where $h_{\max}$ is the maximal deflection of the wafer surface and L the scanning length. Fig. 2.6 shows the curves obtained before and after the deposition and the difference between both curves.

As the film thickness d_f is much lower than the wafer thickness d_s, and the stress is isotropically distributed through the cross section of the film, one can use the Stoney

2.2. STRESS MEASUREMENTS BY SUBSTRATE CURVATURE METHOD

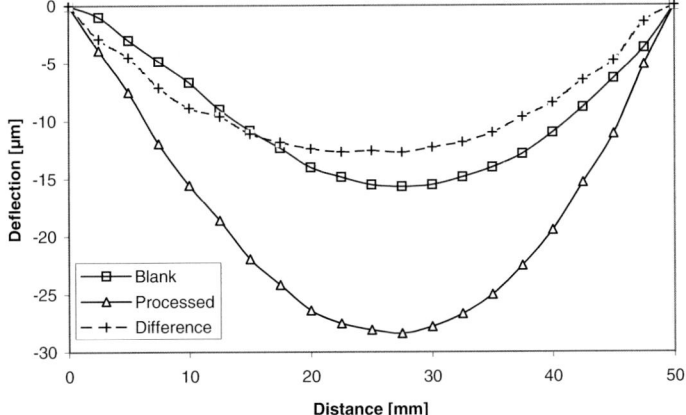

Figure 2.6: Curvature measurements for stress analysis in a stacking of thermal oxide, LPCVD nitride and densified PECVD oxide on a 380 μm thick silicon wafer.

equation taking into account the initial R_{pre} and final curvature radius R_{post} measured before and after the deposition. The second term of the multiplication is now expressed as a radius variation rather than as the radius after deposition.

$$\sigma_f = \frac{E_s d_s^2}{6(1-\nu_s)d_f} \cdot \left(\frac{1}{R_{post}} - \frac{1}{R_{pre}}\right) \qquad (2.32)$$

A maximum deflection variation of around 10 μm was measured on the wafers before the depositions. Following the classical convention, a minus sign was attributed to the radius of curvature for convex wafers (compressive stress) and a positive sign for concave wafers (tensile stress).

Table 2.3 summarizes our results. All values were measured for a starting 380 μm thick silicon substrate using (100) silicon biaxial modulus of 180.5 GPa [60].

Table 2.3: Stress measurement in each layer.

Layer	Layer thickness [nm]	R_{pre} [m]	R_{post} [m]	Stress [MPa]
Thermal oxide	433	90.58	-45.29	**-331**
LPCVD nitride	288	54.16	13.24	**860**
PECVD oxide	283	23.15	57.87	**-397**
Densified PECVD oxide	283	23.15	31.25	**-172**

Stress values agree very well with published data (around -350 MPa for thermal oxide in [27], and between 700 and 1200 MPa for LPCVD nitride in [59]). As we will see in next part, these values will be confirmed by microstructures based measurements. The reproducibility of the method was also confirmed by the fabrication and measurements of independent batches. Thermal oxide and LPCVD nitride showed on other batches larger stresses (between -358 and -400 MPa for the first one and between 888 and 1080 MPa for the second one) always in the literature ranges. Finally delayed measurements do not show any shift (i.e. relaxation) over the time on the stress value at ambient temperature.

A calculation was made using the rigorous Stoney equation without simplification (Eq. 2.26) to check the validity of this simplification. Using values of elastic constants from Table 2.1, a stress of -312 MPa was calculated in thermal oxide, or a difference of 6%. But this error can also be due to incorrect elastic constant values used for the thin film.

It appears that the densification of the PECVD oxide drastically reduces its internal stress. The high value of the stress in PECVD oxide before its densification can be explained by the strong ion bombardments of the growing film by the plasma [59]. High RF power of bombardments results in films with higher density and lower etch rates very attractive in most microelectronics process, but with more compressive stresses. Works of [72] confirm and explain that annealing releases the nitrogen and hydrogen atoms trapped in the oxide and makes the oxide grains grow in size. It proceeds to a rearrangement of the atoms, increasing density of the oxide and decreasing its etch rate and its compressive stress.

The average stress in the previous stacked layers was also extracted. As shown in Table 2.2, sandwiches O-N (thermal oxide and LPCVD nitride) and O-N-O (thermal oxide, LPCVD nitride and densified PECVD oxide) were simultaneously realized following the same process with the previous layers. Measured results can be seen in Table 2.4.

Table 2.4: Stress measurement in stacked layers.

Layer	Layer thickness [nm]	R_{pre} [m]	R_{post} [m]	Stress [MPa]
Sandwich O-N	721	53.33	20.97	**174**
Sandwich O-N-O	1004	30.67	19.53	**80**

Our results indicate that the stacking appears as an interesting method to tailor the average stress and to build 1 μm thick O-N-O membranes with a light tensile stress of

2.2. STRESS MEASUREMENTS BY SUBSTRATE CURVATURE METHOD

80 MPa. This value is a good compromise between surface planarity of the membrane to support all kind of sensors and robustness to be able to move up or down with a maximum of elasticity. It is confirmed by literature: we can read in [73] that from experiments, a thin-film residual stress must be lower than 100 MPa to have good mechanical properties.

Our stoichiometric LPCVD nitride shows a very large tensile stress and needs to be compensated using an oxide. Some other methods were proposed in the litterature to lower down the stress by increasing the silicon proportion in nitride, increasing the SiH_2Cl_2/NH_3 ratio [74][73]. A ratio of four yields **silicon-rich nitride** ($Si_1N_{0.85}$) with a residual stress of around 300 MPa which still constitutes a rather large stress. To decrease further more the residual stress, a three hours post oxidation step ambiance is performed on this above mentioned LPCVD deposited silicon-rich-nitride to tune the film stress to a very low value of 10 MPa without introducing any stress gradient by the newly grown oxide. The paper concludes saying that this method could advantageously replace the sandwich structure (O-N-O) to perform very thin and large membranes (400 nm thick and as large as 4 x 4 cm^2).

Oxynitride films deposited by LPCVD from $SiH_4/N_2O/NH_3$ gaseous mixture also seems to be an interesting alternative in producing very low stressed membranes as demonstrated in [75][76]. The gases ratio can be adjusted to perform a large range of residual stresses and very low stress can be obtained around the stoichiometric relationship Si_2N_2O (or $SiO_{0.5}N_1$).

Nevertheless, when working on SOI (Silicon-on-Insulator) wafers, the 400 nm thick buried oxide is used as the first thermal oxide of our O-N-O stacking. Therefore, the previous methods cannot be interesting and the sandwich O-N-O still remains the best way to produce thin and low stressed membranes on SOI. In this process, nitride will be an added layer replacing the active silicon film and the last PECVD oxide will be the interconnection layer between polysilicon and metal. In the Unibond SOI wafers, the buried oxide is nothing else that a high quality thermal oxide bonded on another silicon wafer. Its residual stress was checked by wafer curvature method. The 100 nm thick silicon film was first etched in TMAH during 1 minute before the measurement of the final curvature radius R_{post}. Silicon oxide (4093 Å) was then etched in HF and the initial curvature R_{pre} was measured. A value of **-363 MPa**, closed to stress measured in our thermal oxide, was finally calculated using Stoney equation taking into account the higher thickness of the SOI wafer (730 μm) in comparison with the thickness of classical bulk wafers (380 μm).

In case of stacked layers, we actually get a stress gradient due to the difference in stress values between the top and the bottom of the stacking. For instance, a stress gradient appears between the top nitride layer and the bottom oxide layer in O-N sandwich. But as explained above, Stoney always measures the average stress in this case. The gradient will be evaluated with dedicated cantilevers and will be widely reported in next section. It will also be explained how stress gradients can have an impact on the robustness of membranes realized from stacked layers.

We predicted the average stress in sandwiches by the superposition Stoney principle from the three isolated thin films. From Eq. 2.28 and Eq. 2.32 we have in case of two stacked films

$$\sigma_{ON} \cdot (d_O + d_N) = \frac{E_s d_s^2}{6(1-\nu_s)} \cdot \left[\left(\frac{1}{R_{O\,post}} - \frac{1}{R_{O\,pre}}\right) + \left(\frac{1}{R_{N\,post}} - \frac{1}{R_{N\,pre}}\right)\right] \quad (2.33)$$
$$= \sigma_O d_O + \sigma_N d_N$$

where σ_{ON} is the average stress in the sandwich structure O-N of thickness $(d_O + d_N)$, σ_O, the average stress in the oxide layer of thickness d_O, and σ_N, the average stress in the nitride layer of thickness d_N.

Using this previous equation, values of **144 MPa** and **55 MPa** are calculated respectively for O-N and O-N-O multilayers from the curvature values of the monolayers. The differences between the measured and calculated values are explained using the same Stoney superposition principle but starting from the sandwiches and decomposing them into isolated layers (by removing the partial stress from the total stress). Furthermore, this calculation will lead to some interesting conclusions about thermal behavior.

If the substrate curvature due to the stress in the densified PECVD oxide is substracted from the substrate curvature due to the average stress in the sandwich O-N-O, a value of **179 MPa** is obtained for the stress in the sandwich O-N, very close to the measured value (**174 MPa**). It confirms that the deposition at 300°C of the PECVD oxide followed by a densification at 800°C does not affect the stress in the sandwich O-N underneath. It could be predicted since LPCVD nitride and thermal oxide were obtained respectively at 1000°C and 800°C.

Similarly, by removing the stress caused by the nitride layer from the stress of the sandwich O-N, one obtains the final stress in the thermal oxide. -281 MPa is calculated which is not so close to the -331 MPa directly measured in the oxide film. It could mean

2.2. STRESS MEASUREMENTS BY SUBSTRATE CURVATURE METHOD 67

that LPCVD nitride deposition affects the thermal oxide layer and removes a small part of its residual stress. To verify that, we put our oxidized wafer in the LPCVD oven to simulate nitride deposition at the same temperature during same time but without gas. The curvature measurements at the end of this test gave an average stress of -319 MPa. Temperature deposition of LPCVD nitride does therefore practically not affect the stress in thermal oxide layer as it could be expected since oxide was grown at a higher temperature than nitride. Temperature can therefore not be the reason for the decrease when nitride is actually deposited. SiH_2Cl_2 and NH_3 gases penetrating inside the oxide layer during the nitride deposition could maybe modify its internal structure and decrease its residual stress. More consistently, LPCVD nitride grown on silicon oxide seems to have a higher residual stress than LPCVD nitride grown on silicon due mainly to the difference in numbers of nucleation[3] sites between silicon surface and silicon oxide surface. As there are more nuclei growing on amorphous oxide than on cristalline silicon, resulting nitride grains are wider on silicon than on silicon oxide. And since stress is inversely proportional to the grains size [59], the stress in nitride layer deposited on silicon oxide is increased. So residual average stress in nitride deposited on oxide becomes therefore **914 MPa** instead of the lower value of 858 MPa. Nevertheless, the difference caused by the measurement of the nitride stress on substrate is only 6 % thanks to its already very high stress value.

This difference of stress in the LPCVD nitride layer deposited on oxide instead of silicon explains the difference between measured and calculated values of stress in case of sandwichs O-N and O-N-O. Taking into account the non negligible effect of the nucleation, the Stoney superposition principle has therefore demonstrated its ability to predict the average stress of a stack when the average stress of each constitutive layer is known.

The curvature method was finally used to measure the final average residual stress in a O-N-O membrane at the end of a complete gas sensor process. To perform that, the curvature of a wafer undergoing the process steps was measured before and after the process. The other layers besides the constitutive layers of the membrane were removed just after their deposition. The idea was to check if steps like polysilicon and aluminum deposition, annealing and etching could affect the average stress of a membrane. In this process, thermal oxide, LPCVD nitride and densified PECVD oxide were obtained in the same way as explained above but their thicknesses were slightly different: 430 nm for thermal oxide, 340 nm for LPCVD nitride and 390 nm for PECVD oxide. Previous values

---

[3]Nucleation can be defined as the initial stage in a phase transformation, evidenced by the formation of small particles (nuclei) of the new phase, which are capable of growing. Nucleation is always depending on the substrate surface [27].

of stress with the new thicknesses of these layers can be used to predict the final average stress using the superposition principle of Stoney. The thickness variation would have no impact on the previous stress values of LPCVD nitride and PECVD oxide since these layers are uniform through their thickness. It is a priori not the case for thermal oxide but its thickness is exactly the same than in our previous tests. Therefore, by Stoney superposition principle using stress values in Table 2.3 and new thicknesses, we foresaw an average stress around 100 MPa. 160 MPa were measured by wafer curvature on the test wafer at the end of the process. To explain this higher tensile stress, layers were removed one after the others, curvature measurements were performed between each etch and stress values were calculated using the superposition principle of Stoney. -174 MPa for densified PECVD oxide, 1027 MPa for LPCVD nitride and -332 MPa for thermal oxide were extracted by this way. These results reveal therefore that PECVD oxide and thermal oxide were not affected by the process given their values very close to the previous ones (see Table 2.3). Only the stress in LPCVD nitride, once again, was increased in comparison with the 915 MPa extracted above, giving an underestimation of the final tensile stress.

2.2.3 Discussions and summary

Substrate curvature method seems to yield a simple and fair estimation of the average stress in the three constitutive layers of our membrane as well as in their stacking. A low tensile average stress of 80 MPa was extracted in a O-N-O membrane and was found very close to the predicted value calculated using Stoney superposition principle on each isolated layers. By the decomposition of the sandwich to each isolated layers, we explained the physical behaviors at the origin of the differences between measured and calculated values. This method was finally validated by comparing the final measured average stress of the three stacked layers of our membrane at the end of our gas sensor process with the calculated expected value. It revealed that curvature measurement and its superposition principle can be very attractive to foresee the mechanical properties of multilayers.

Advantages and drawbacks of these methods can finally be summarized as following:
Advantages of curvature measurement method:

- Requires only fast and easy measurements,

- Does not need any photolithographic step,

- Gives the average stress of a thin film without a priori knowledge of its elastic properties (Young's modulus and Poisson ratio),

2.3. STRAIN MEASUREMENTS USING MICROMACHINED STRUCTURES

- Provides an average stress including the effect of the stress gradient,

- Can be used to predict the average stress of staked layers from the knowledge of the average stress and the thickness of each of them.

Drawbacks of curvature measurement method:

- Provides the average stress value over the entire test wafer and does not take into account the local fluctuations on the wafer and the fluctuations from one wafer to another (since measurements are performed on a test wafer and not on the actual processed wafer),

- Requires elastic properties of the thin film under consideration to extract the average strain.

- Provides an average stress integrated over the whole thickness of a layer but does not quantify the stress gradient in it.

To complete this detailed study on the average residual stress measurements by wafer curvature measurement technique, it was needed to check if a complementary method compensating these drawbacks, could yield comparable results. The presentation and development of this complementary technique are the focus of the next section.

2.3 Strain measurements using micromachined structures

As explained above, substrate curvature measurements result in an average wafer level value of the stress rather than local values. Local stress does not necessarily mean the same as stress measured by substrate bending techniques, since stress is defined microscopically, while deformations are mostly induced macroscopically [27]. Residual stress can vary across a wafer but also from a test wafer to a processed wafer and from lot to lot. It is therefore essential to quickly accurately access local residual stress values during and after wafer processing. Local measurements have also the advantages that measurements are made on the same dimensional scale as the film of interest and that the local stress field can be mapped [77].

The local film strain can be measured using in situ micromachined structures made directly out of the film of interest itself. When a thin film structure under stress is released by removing the underlying silicon, the residual stress in this structure is relaxed and is converted into a measurable increase or decrease of the structure dimensions. This deformation is the relaxed strain proportional to the stress included in the thin film before its relaxation. The proportionality factor in this case is the uniaxial Young's modulus since each microstructure presented in this section are characterized by a length much longer than its width [78][79][65]. Moreover, our microstructures are subject to uniaxial compression or traction and the beam sides are free for most of the length. Young's modulus is therefore required to extract the final average stress from the measured strain if necessary. But on the contrary to the curvature measurement method, the microstructures presented in this section yield measurements of the strain without the knowledge of the elastic constants of the silicon substrate and the measured thin film.

Nevertheless, both methods will be compared since the elastic constants of the silicon and the three dielectric layers in consideration are known with good accuracy. But the strain of a layer having unknown elastic constants would be advantageously extracted from microstructures and the stress from curvature method to finally calculate the Young modulus and Poisson ratio of this layer for instance. Both methods remain therefore complementary. To perform accurate comparisons, in the following, each local film stress measurements was performed on layers deposited during the same process than for previous curvature measurements (on same wafers than wafers in Table 2.2).

The change in length when the microstructure is relaxed is however very small, making direct measurement difficult. It is therefore necessary to convert this change into buckling of a part of the structure or into a larger displacement. The buckling of a clamped-clamped beam appears above a critical compressive stress. The stress is estimated from the maximum length possible without buckling and requires therefore an array of devices with incrementally increasing lengths. In another way, to increase a displacement we can transform the extension or contraction of a supported beam due to the strain in a rotation of a second beam. The deflection of the tip of this beam will be a few times larger than the extension or contraction of the primary beams and is an indication of the strain value in the structural layer. These structures will be presented in next paragraphs.

While contact profilometer was used to perform substrate curvature measurements in our lab[4], only nocontact optical methods are necessary to measure the strain in released

---

[4]Note that substrate curvature can be measured by no contact methods such as laser scanning.

2.3. STRAIN MEASUREMENTS USING MICROMACHINED STRUCTURES

structures. Scanning electron microscopy (SEM) and optical microscopy were used for in-plane measurements. But for thin films, normal mode of buckling is out-of-plane. Optical microscopy can be used to perform that but it requires focal plane adjustments which are not really accurate for small deflections. Microscopic interferometry [80] is a more sensitive technique for detecting small buckling. A home made interferometric setup based on a low coherence light emitting diode (LED) source was used [81][82] and was seen powerful. The principle of this technique is detailed in Annex B with illustrations acquired during our measurements.

Finally, it could be interesting to note that in our case, the micromachining is at the middle way between surface and bulk micromachining since silicon bulk is partially etched to release the thin films structures. All structures were especially designed to take into account the TMAH silicon etching property, i.e. the undercutting of masking material at convex corners (see previous Chapter).

2.3.1 Clamped-clamped beam and ring-and-beam structures analysis

Theory

A doubly supported beam (Fig. 2.7) under compressive stress will buckle when it is released. An array of this kind of microstructures with various lengths can be therefore used to estimate a range of **compressive strains** by determining which of them have buckled at a given stress level [11]. The critical strain ε_{cr} can be estimated in this case using the Euler equation for a critical buckling beam of length L_{cr}

$$\varepsilon_{cr} = \frac{\pi^2 h^2}{3 L_{cr}^2} \tag{2.34}$$

where h is the beam thickness, and L_{cr}, the critical beam length (i.e. the shorter beam at which buckling occurs). So, for a given layer thickness, a higher stress leads to a shorter critical buckling beam and reverse. However, if the layer thickness h decreases, the critical buckling length L_{cr} decreases in the same way.

Furthermore, as the amplitude of the buckling is sensitive to residual stress in the film, this allows each buckled beam to be used for residual strain measurement ε_C in the

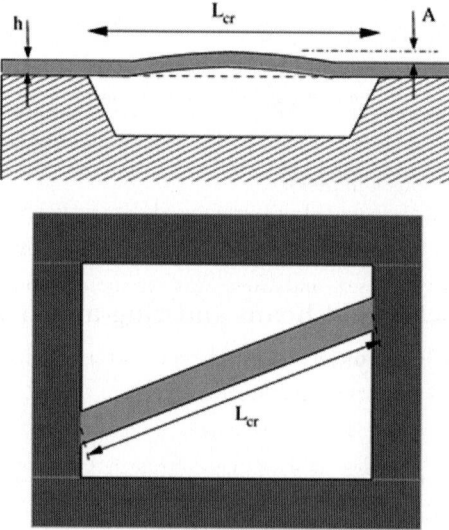

Figure 2.7: Cross view and top view of a clamped-clamped beam released on silicon. The beam was drawn obliquely across the silicon cavity to be released by silicon anisotropic etching (see later).

2.3. STRAIN MEASUREMENTS USING MICROMACHINED STRUCTURES

post-buckling regime [54]. Residual strain for each buckled beam can be determined from

$$\varepsilon_C = \frac{\pi^2}{12L^2}\left(3A^2 + 4h^2\right) \qquad (2.35)$$

as demonstrated in details in [54] and [78] where A is the amplitude of the buckling deflection, L is the beam length and h is the beam thickness. The shape of the buckling beam is sinusoidal as exposed by [54]. Both A and h can be accurately measured by interferometry and ellipsometry respectively, and L is well-known from the layout dimensions. Eq. 2.35 gives excellent results for the residual strain measurement using beam of lengths somewhat larger than the ideal critical Euler buckling length L_{cr} [54]. We will see further that it is very interesting in our case where films thicknesses are very thin showing very short buckling lengths for a given critical strain and therefore difficult to be detected by optical interferometry.

To accurately extract the effect of the strain gradient on clamped-clamped beam deflection, the full deflection curve was compared in [54] with a two-dimensional finite difference model. It was found from these simulations that for the post-buckled beams, use of Eq. 2.35 rather than the more complete model results in less than a 3% difference in the residual strain results. This indicates that the effect of the gradient on the residual strain measurements is small on post-buckled beams. The further going into the post-buckled regime, the less the gradient affects the buckled geometry and therefore the measured residual strain. On the other hand, the model shows that gradient affects the evaluation of the transition and prebuckled beams [54]. It is one more motivation to use the post-buckling measurement in case of thin film thermal oxide for instance since it contains internal strain gradient.

A microstructure under residual tensile stress is more difficult to be relaxed when it is released to express its **tensile strain**. So, the idea was therefore to design a structure transforming the tensile stress into a compressive stress which could lead to a measurable deformation or strain. The structure performing that is the ring as shown on Fig. 2.8. In this structure, a circular ring anchored at two diametrically opposite positions converts tensile residual stress into compressive stress in a crossbeam that is orthogonal to the ring anchors [54]. A tensile stress in the layer deforms the ring into an ellipsoid, and the stress in the central beam becomes compressive. Ranges (rather than precise measurements) of residual strain levels are determined by observing the transition from unbuckled to buckled crossbeams as for clamped clamped beams.

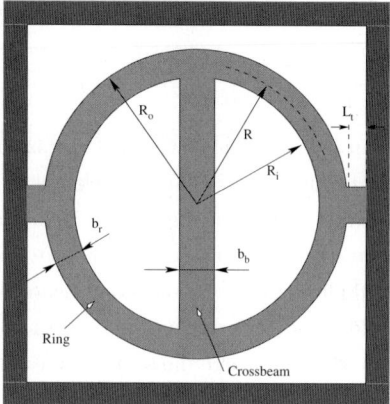

Figure 2.8: Top view of a ring-and-beam structure ready to be released on top of the silicon cavity.

The residual strain for the structure is

$$\varepsilon_R = \frac{\varepsilon_{cr}}{g(R_{cr})} \tag{2.36}$$

where ε_{cr} is the critical strain needed to buckle the crossbeam and $g(R_{cr})$ is the conversion efficiency of tensile strain in the first buckled ring into compressive strain in the crossbeam.

The critical strain ε_{cr} can be approximated to be the same as the critical buckling strain in the clamped-clamped beam from Eq. 2.34 where the length of the beam L_{cr} was expressed in terms of ring radius ($L_{cr} = 2R_i$). This is not completely true since the crossbeam or buckling beam in the ring structure is not clamped at the ends. The ends will even move due to the deformation of the ring when it converts the tensile strain in the layer into a compressive strain on the beam [11]. [83] calculates the exact value of ε_{cr} taking into account the out-of-plane torsion of the ring. With the previous approximation, Eq. 2.36 yields:

$$\varepsilon_R = \frac{\pi^2 h^2}{12 \cdot R_i^2 \cdot g(R_{cr})} \tag{2.37}$$

2.3. STRAIN MEASUREMENTS USING MICROMACHINED STRUCTURES

The conversion efficiency $g(R_{cr})$, is found from the following equations [83] [54]:

$$g(R_{cr}) = \frac{-(2b_r f_2)}{(2b_r f_1 + b_b f_1^2 - b_b f_2^2)} \quad (2.38)$$

$$f_1 = \left(\frac{\pi}{4} - \frac{2}{\pi}\right)\left(\frac{R_{cr}}{e}\right) - \frac{2e}{\pi R_{cr}} + \frac{4}{\pi} - \frac{\pi}{4} + \frac{\pi k_f (1+\nu)}{2} \quad (2.39)$$

$$f_2 = \left(\frac{1}{2} - \frac{2}{\pi}\right)\left(\frac{R_{cr}}{e}\right) - \frac{2e}{\pi R_{cr}} - \frac{1}{2} + \frac{4}{\pi} - k_f(1+\nu) \quad (2.40)$$

Definitions of the geometrical parameters b_r and b_b can be found in Fig. 2.8. The constant k_f is a form factor for transverse shear ($k_f = 1.2$ for rectangular cross sections [83]), ν is the Poisson Ratio, R_{cr} is the effective ring radius, and e is the eccentricity,

$$e = R_{cr} - \frac{b_r}{\ln\left(\frac{R_o}{R_i}\right)} \quad (2.41)$$

with

$$R_{cr} = \frac{(R_o + R_i)}{2} \quad (2.42)$$

$$R_i = R_{cr} - \frac{b_r}{2} \quad (2.43)$$

$$R_o = R_{cr} + \frac{b_r}{2} \quad (2.44)$$

With a scaled array of ring-and-beam test structures, the approximate value of the residual tensile film stress can therefore be determined by observing for which length crossbeams have buckled.

It seems important to note that the strain gradients affect the bending of microrings, meaning that the entire shape and not only the crossbeam flexures must be measured in order to correctly interpret the deformation of the structure [54]. In fact, a tensile state would only be indicated by significantly different out-of-plane flexures for the crossbeam compared to the ring. Without the complementary information from the cantilevers to detect any strain gradient, we could incorrectly conclude that if the crossbeams were buckled, the residual strain in the film was tensile. But as we will see further, in our case, tensile LPCVD nitride does not show any gradient which could be at the origin of erroneous measurements.

Experimental setups and results

As it has been mentioned in the previous theoretical part, in case of our two buckling structures, an array of structures is needed to be able to measure strains within a defined range.

From literature and from our previous results using wafer curvature method, we know that the compressive stress in thermal or PECVD oxide can vary around -350 MPa. From its well known Young's modulus of 70 GPa and using the Hooks law (Eq. 2.6), the proportional strain range was evaluated between 0.1 and 1 %. Equivalent critical lengths were calculated using Eq. 2.34 in the range of 7 to 23 μm (using $h = 433\ nm$ from Table 2.3). However, longer beams were designed to make measurements in post-buckling state until 50 μm. All beam lengths in the array have been carried out in several beam widths of 3, 5 and 8 μm. Theoretically, the beam width has no effect on critical strain (it does not appear in Eq. 2.34) but in practice, it is not really the case.

In the same way, tensile stress values in stoichiometric LPCVD nitride are generally around 1000 MPa, either a strain value around 0.37 % (with a Young's modulus of 270 GPa). An array was designed around this value of strain from 0.1 to 1 % for three ring-and-beam width ratio (i.e. b_r/b_b ratio): 3/5, 4/2 and 3/2. From Eq. 2.37 and following (with $\nu = 0.27$ and $h = 288\ nm$), the ring radius R was therefore calculated in the range of 7 to 40 μm. The tie length L_t was fixed equal to 0.5 or 1 μm (depending on the ring size) as close to zero as possible since equations were designed without ties. Finally, for drawing the masks, rings were approximated in polygons by dividing the circle into 22 straight lines.

Tables 2.5 and 2.6 summarize both designed arrays.

Table 2.5: Clamped-clamped beam array.

width [μm]	first length L [μm]	last length L [μm]	step [μm]	#
3	7	50	0.70	62
5	7	50	0.70	62
8	7	50	0.70	62

As silicon is used as sacrificial layer, it was needed to take into account the properties of anisotropic silicon micromachining by TMAH for their design. Thanks to its property

2.3. STRAIN MEASUREMENTS USING MICROMACHINED STRUCTURES

Table 2.6: Ring-and-beam array.

width ratio b_r/b_b	first radius R [μm]	last radius R [μm]	step [μm]	#
5/3	7.5	11.5	0.1	57
	11.5	15	0.5	
	15	20	1	
	20	40	5	
4/2	7	11.5	0.1	62
	11.5	15	0.5	
	15	20	1	
	20	40	5	
3/2	6.5	10	0.1	50
	10	12	0.5	
	12	20	1	
	20	40	10	

of undercutting of masking material at convex corners (see previous chapter), clamped-clamped beams were therefore drawn obliquely at 45° across the cavity to be released without damaging the anchors (as shown of Fig. 2.7). As it will be demonstrated further, this special geometry could slightly underestimate the measured strain but Eq. 2.34 and 2.35 always remain valid. In this case, length was measured between both anchors, at the middle of the beam as shown in Fig. 2.7 to give an average length.

On the other hand, rings do not need any special configuration because their intrinsic geometry involves convex angles allowing the complete structure to be released in TMAH without damaging the anchors.

The fabrication process to release our structures was in this case easier than a classical process to release polysilicon structures on sacrificial layer. Only one mask and one photolithography step are needed. After growing the thin film on the silicon substrate, structures were defined by photolithography and patterned using the appropriate etchant (plasma for nitride and PECVD oxide, HF for thermal oxide). Wafers were then put in HF during 10 seconds to remove native oxide on silicon before TMAH etching. Silicon etching was performed in a TMAH 20% solution at 90°C just long enough to release the complete structures; i.e. during one hour, providing a cavity depth of around 60 μm. Wafers were finally rinsed in DI water and dried in methanol in place of classical rinser dryer to avoid structures damage.

Buckling detection was first performed using focal plane adjustments of optical microscopy to give an approximate value of average stress in our thin films. But this tech-

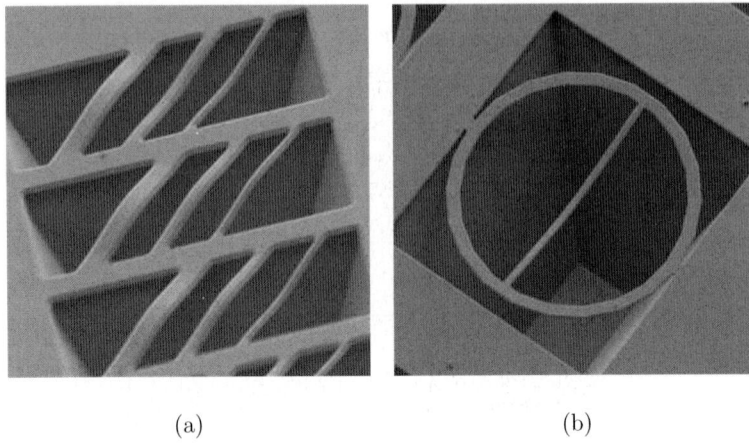

Figure 2.9: (a) SEM view of the last nine clamped-clamped beams of the array (as defined in Tables 2.5 and 2.6) in thermal oxide and (b) last LPCVD nitride ring-and-beam (radius $R = 40$ μm, $b_r = 4$ μm and $b_b = 2$ μm).

nique was demonstrated to be not so relevant due to the really small dimensions of our beams. In order to increase their dimensions and therefore their visibility, the solution could be to increase their thickness. Nevertheless, thicker layers are difficult to obtain and would lead to a different internal structure (due to their intrinsic stress for instance) and therefore, not to the same average stress. Interferometry was therefore preferred as shown in Fig. 2.10 and 2.11.

In the case of **thermal oxide**, the detection of the critical buckling beam was impossible. The very low thickness indeed resulted in a too small critical beam for a high residual stress as shown in Eq. 2.34. However, we observed a decrease of the buckling at the end of the array, i.e. for stresses around -400 MPa. Moreover, oxide has an intrinsic gradient which could alter the detection of the critical buckling beam as explained above. Therefore, measurements were performed on post-buckled beams by interferometry from the 12th to the 62nd beams for each of the three widths to analyze the effects of the length and the width. Obtained profiles can be seen on Fig. 2.12.

Strain was calculated from the amplitude of the deflection using Eq. 2.35 and stress was extracted using the Young's modulus of thermal oxide (70 GPa). Variations of calculated stress versus the beam lengths and widths are depicted at the top of the graphic in Fig. 2.13 and yield an average stress value between -280 and -330 MPa.

2.3. STRAIN MEASUREMENTS USING MICROMACHINED STRUCTURES 79

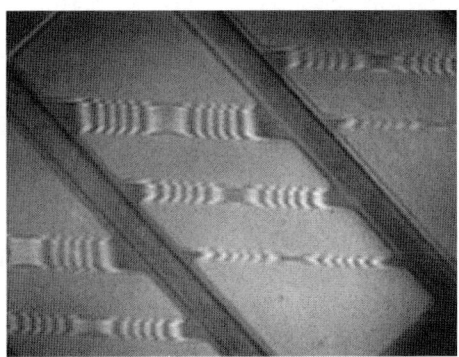

Figure 2.10: Interferogram of thermal oxide 50 μm long clamped-clamped beams.

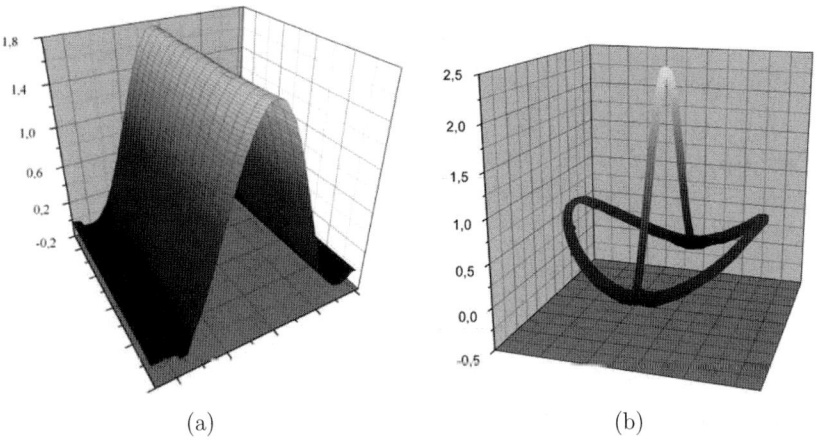

Figure 2.11: Tridimensional view of the deflection (in μm) extracted by interferometry: (a) clamped-clamped beam and (b) ring-and-beam structure.

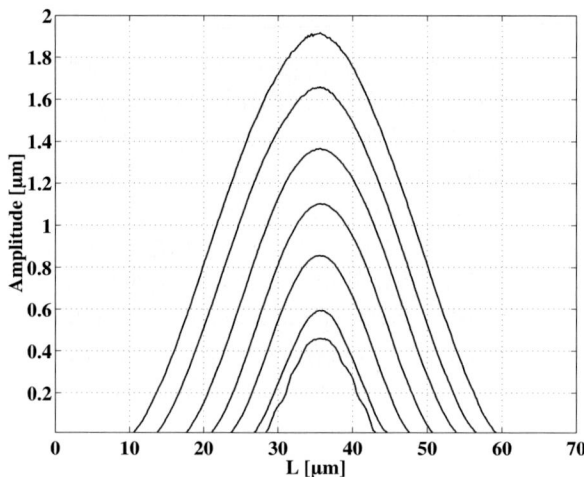

Figure 2.12: Thermal oxide clamped-clamped beams deflection for one width of 5 μm and 7 different lenghts: 50.20, 44.55, 37.48, 31.11, 25.46, 19.80, 17.68 μm.

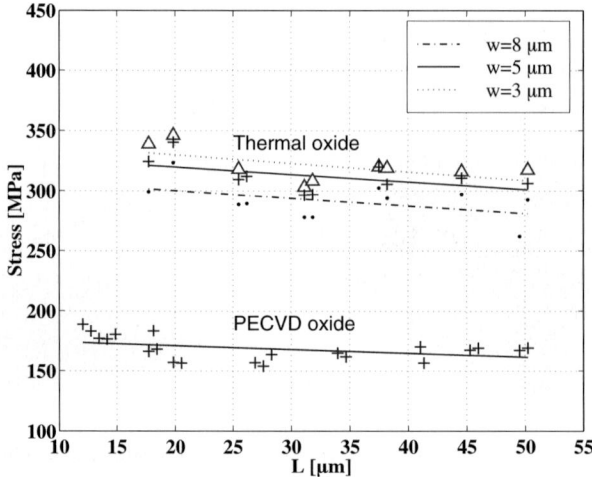

Figure 2.13: Measured stress on clamped-clamped beam array vs. length of the beam for 3 different widths in case of thermal oxide and for a beam width of 5 μm in case of PECVD oxide.

2.3. STRAIN MEASUREMENTS USING MICROMACHINED STRUCTURES 81

Stress appears slowly decreasing when the length increases. This could be due to the fact that the stress gradient has a large influence on the buckling of shorter beams and less effect as the beam gets longer. Regarding the difference of stress versus the width of the beam, a wider beam shows a lower stress than a narrow beam. We will make the same analysis on cantilevers at the end of this chapter, i.e. wider beams have a greater stiffness and would therefore be less relaxed than narrower beams. The geometry of the beam clamped obliquely also provides a little more rigidity to the beam near their anchors. It was confirmed by ABAQUS mechanical simulations: a higher stress concentration was observed at the anchors. This effect decreases of course with the beam width and in borderline case, it vanishes when the beam width is close to zero. So, the medium width (5 μm) seems a good trade-off between medium stiffness, great measurement facility and low error induced by the geometry. Finally, measurements of 5 μm wide beams on other sites showed an average range of stress values (for beams of 35 μm long) between **-300** and **-330 MPa**.

Compressive strain in **densified PECVD oxide** was also characterized using clamped-clamped beams array. As the strain measurement is depending on the thickness of the layer, we took into account the fact that PECVD oxide was attacked significantly in TMAH on the contrary of thermal oxide and LPCVD nitride (see previous chapter). A thickness of 275 nm was measured after one hour of etching giving also an expected critical buckling beam impossible to be detected. Measurements were therefore performed by interferometry on 5 μm wide post-buckled beams and gave an average stress between **-160** and **-190 MPa** as shown at the bottom in Fig. 2.13.

The detection of the critical buckling crossbeam in the array of rings was finally performed by interferometry to measure the tensile stress in **LPCVD nitride**. First buckling crossbeam was initially impossible to detect since each crossbeam and each ring of the array appeared buckled. But measurements of the deflection height on three points of the ring-and-beam structure (on the anchor, at one extremity of the crossbeam and at the middle of the crossbeam) revealed an interesting behavior: the deflection of the crossbeam had its sign changing once accross the array. Therefore, the critical buckling was extracted by detecting the sign change in the crossbeam deflection. The absolute deflection of the crossbeam was extracted removing from its value the heights of the anchor and of one extremity of the crossbeam. The graphic in Fig. 2.14 shows the evolution of the absolute deflection of the crossbeam versus the critical stress calculated for each ring from its ring radius (with a Young's modulus of 270 GPa). The second curve shows this proportionality between the calculated critical stress and each ring radius. It clearly depicts that the crossbeam deflection changes sign in this case around 990 MPa.

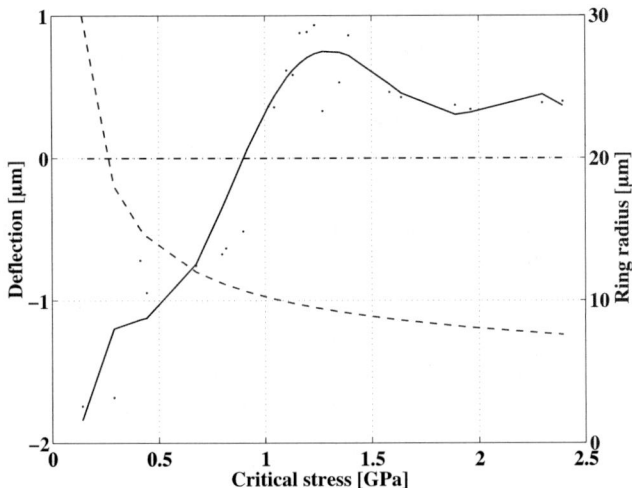

Figure 2.14: Absolute deflection of the crossbeam versus critical stress calculated for each ring from its ring radius (second curve).

Detection therefore became very easy to perform. This way, measurements were carried out over the wafer and showed stresses in the range from **840** to **1015** MPa with a higher repeatability of values on the left of this range. The best reproducibility was reached for the 4/2 width ratio showing the higher difference between the beam and the ring width. The measurement also showed a large difference in buckling amplitude between the ring and the crossbeam demonstrating that no gradient was present [54]. This will be demonstrated further using cantilevers.

2.3.2 Microgauges analysis

Both clamped-clamped beams and ring-and-beam structures need to be implemented in entire arrays of structures and they could not be so easily integrated with active microstructures due to space constraints. In contrast to previous structures, only one microgauge is sufficient to determine any **tensile or compressive strain** under optical microscope independently on the deposited thin film thickness [55][27].

2.3. STRAIN MEASUREMENTS USING MICROMACHINED STRUCTURES

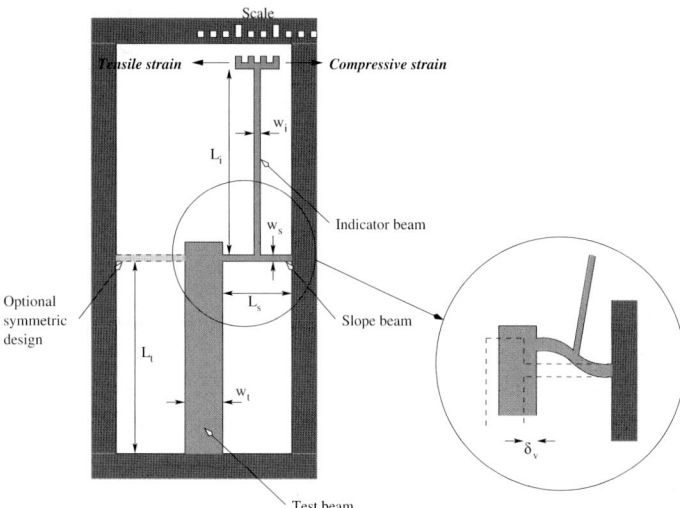

Figure 2.15: Microgauge designed to be released on silicon and zoom of the slope beam when a high compressive stress occurs.

As shown in Fig. 2.15, the microgauge consists of three beams, a test beam, a slope beam, and an indicator beam. All of them are suspended above the substrate to move freely but are anchored at two points, one at the end of the test beam and another at one end of the slope beam. Residual strain existing in the thin film causes the test beam to either expand (under compressive residual strain) or contract (under tensile residual strain). This movement is transferred to the slope beam causing its deflection into an "s" shape (see Fig. 2.15) since the other end of the slope beam is a fixed anchor. The indicator beam, which is placed at one end at the center of the slope beam, amplifies this tiny rotation in a large displacement of its other end where is placed a scale (or a vernier) to simplify the reading of the deflection under microscope. A compressive stress moves the indicator beam to the right due to the elongation of the test beam and a tensile stress leads to a deflection to the left. Fig. 2.15 depicts that a second slope beam can be put on the opposite side of the test beam with another anchor. This symmetrical design shows the same strain readout than the one-sided design [55] but the added anchor gives more rigidity to the structures fabricated in very thin films. A disadvantage of this modified structure is that it buckles easier since residual stress in the slope beam cannot be relaxed.

The residual strain in the test beam measured by the deflection of the indicator is demonstrated in details in [55] and can be represented as

$$\varepsilon_g = \frac{2L_{sb}\delta_v}{3L_{ib}L_{tb}C} \qquad (2.45)$$

where L_{sb}, L_{ib} and L_{tb} are respectively the lengths of the slope beam, the indicator beam and the test beam, respectively. δ_v is the measured deflection read from the gauge scale. C is a correction factor due to the presence of the indicator beam and is derived as

$$C = \frac{1-d^2}{1-d^3} \quad \text{with} \quad d = \frac{w_{ib}}{L_{sb}} \qquad (2.46)$$

where w_{ib} is the width of the indicator beam.

The accuracy of the microgauge is greatly improved because its output is independent on the film thickness but also on the cross-section of the microstructure which could have a trapezoidal shape after a wet etching. Finally, the strain measurement with this kind of microstructure is not affected by the out-of-plane strain gradient. However, in presence of high intrinsic strain gradient, the movable part of the indicator bends upwards, and it is difficult to focus the microscope to both scales simultaneously.

Similarly to clamped-clamped beams and rings, the gauges were released by etching silicon underneath with TMAH. As the structure involves a lot of convex corners, a rectangular cavity was straightforwardly etched under the structure, three sides providing the anchors and the fourth side providing the reference scale as shown by the SEM picture in Fig. 2.16. The etching time was more critical to control in this case to avoid undercutting of the anchors during beams release. One hour was shown optimal and yielded sufficiently deep cavities (around 60 µm) to avoid any stiction problem.

In our case, no vernier was integrated on the tip of the indicator beam as it is usually performed in gauges [55] but only a scale corresponding to another reference scale on silicon to be able to measure the displacement by SEM. It was easier to design and more suitable in our case of very thin dielectric films. Due to the very low thicknesses of our layers, in spite of an accurate design, the vernier would be deformed and laterally etched, leading to incorrect reading of the strain.

Several dimensions were tested but only one gauge was completely released after one hour of etching and gave good results. Its dimensions are summarized in Table 2.7. They result of a trade-off between high precision in the measured strain range, great stiffness of the structure after its etching and complete release of the three beams in a short time

2.3. STRAIN MEASUREMENTS USING MICROMACHINED STRUCTURES

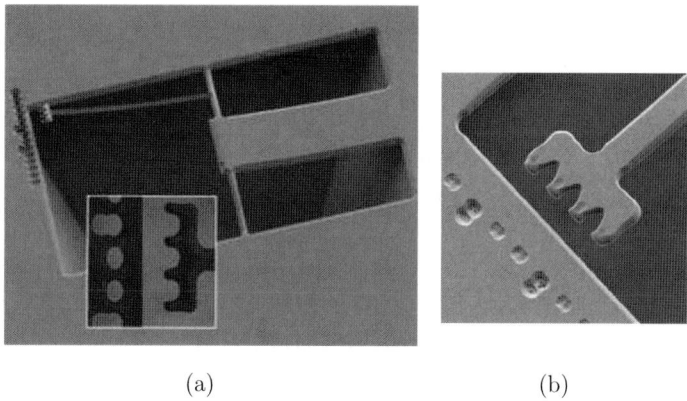

(a) (b)

Figure 2.16: (a) SEM view of a gauge in thermal oxide and (b) zoom of the tip of a LPCVD nitride gauge (for dimensions, see Table 2.7).

without damaging the anchors. Moreover, the notches at the top of the indicator beam scale and on the reference scale were chosen with a width of 1 μm and a separation of 1.5 μm considering the limitations of the optical photolithography.

Table 2.7: Design values of the microgauge.

| $\mathbf{L}_{tb}$ | $\mathbf{L}_{ib}$ | $\mathbf{L}_{sb}$ | $\mathbf{w}_{tb}$ | $\mathbf{w}_{ib}$ | $\mathbf{w}_{sb}$ |
[μm]	[μm]	[μm]	[μm]	[μm]	[μm]
53	50	20	20	2	2

Displacements of the indicator beam were measured using SEM imaging. An optical microscope in its maximum magnification (100X objective) would be sufficient but even without gradient (in case of LPCVD nitride), the thin indicator beam tended to buckle a little out of plane giving large out of focus problems. Measurements were achieved on several sites to determine a range of stress instead of a constant value. Thermal oxide exhibited deflections from 1.20 to 1.30 μm rightwards (measured by SEM) which reported in Eq. 2.45 and 2.46 lead to values of stress from **-460** to **-490 MPa** (with Young's modulus of 70 GPa). Indicator beam in LPCVD nitride was deflected to the left from 0.58 to 0.70 μm showing a tensile stress from **795** to **950 MPa** (with Young's modulus of 270 GPa). Gauges in PECVD oxide did not give reproducible results due to their damaging during TMAH etching. When the thin beam was released during etching, the beam continued to be damaged from the top as from the bottom and the side walls of the beam. So, the thickness and the width of the indicator beam decreased slowly during etching and finally tended to be a useless wire distorted in all directions.

An alternative interesting microstructure derived from the microgauge is proposed by [56]. It consists of a pair of two cantilever test beams with different lengths connected by a short indicator tip beam. The difference of contraction or elongation between the two test beams under strain causes the deflection of the indicator beam. The errors in displacement measurement due to the stress gradient or the nonvertical side walls (which can appear after some wet isotropic etching) effects are demonstrated negligible. As shown in [56], the advantage of this structure in comparison with the classical microgauge is that it does not need the slope beam to amplify the movement and therefore gives measurements without the errors it introduces. A symmetrical design is also presented to measure a double displacement. This structure was tried with success in our lab to make strain evaluation in 2 μm thick polysilicon layers using a sacrificial oxide layer. Nevertheless, its geometry is less compatible with the anisotropic wet etching to release the underlying silicon substrate without damaging the anchors and this method was therefore not chosen for the present purpose.

2.3.3 Cantilever beams analysis

Monolayer cantilever beams

Previous microstructures were shown useful to measure uniform compressive or tensile strain. We studied the limitations of these structures when some non uniformities or stress gradients occur through the depth. It is therefore very important to measure if a stress gradient exists in the structural layer to verify the validity of our measurements with the previous microstructures. Moreover, it is interesting to see if changing some parameters in the process steps of the layer deposition could result in a smaller intrinsic stress gradient. The cantilever will allow us to perform this measurement as it will be demonstrated below.

Fig. 2.17 shows a cantilever beam clamped at one side, built in a thin film (thickness h) with an intrinsic stress gradient, deposited on silicon substrate. Before the cantilever release, there is both an average stress (compression) and a stress gradient in the thin film. The average stress is a thermal mismatch stress between the thin film and the substrate and the stress gradient is an intrinsic stress in the thin film. If we suppose the film is released but constrained to stay flat, the average stress would go to zero since

2.3. STRAIN MEASUREMENTS USING MICROMACHINED STRUCTURES 87

the film would be free to expand or contract. The uniform stress would in this case be replaced by a uniform relaxed strain. But the film would stay in tensile stress on top and compressive stress at the bottom and therefore in stress gradient. Without such constraints, the freestanding film will curve towards the side that is in tensile (positive) stress until the stress is relaxed. The stress gradient is then vanished and replaced by a relaxed strain gradient. The curling of the cantilever can therefore be used to measure this relaxed strain gradient.

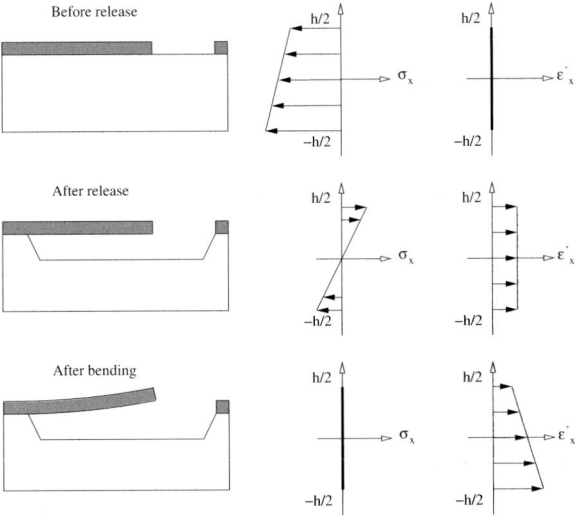

Figure 2.17: Curvature caused by stress gradient in freestanding thin film of thickness h. Diagram of approximated stress and relaxed strain $\varepsilon\prime$ (as defined above) through the thickness.

We can consider the radius of curvature as a function of strain gradient or more precisely as a function of stress gradient since the cantilever will curve from the constraint state (under stress gradient) to the completed relaxed state as shown in Fig. 2.17. This is a particularly simple case in which the treatment of a cantilever beam exactly follows that of a plate (as demonstrated at p. 56), assuming that the biaxial modulus needs to be replaced by the uniaxial modulus. From Eq. 2.20 calculating the maximum stress on top and at the bottom of the film, we can therefore extract the stress gradient $\Delta\sigma$ (or the strain gradient $\Delta\varepsilon$) as the difference between the stress at the top ($+\sigma_m$) and bottom

$(-\sigma_m)$ assuming that the stress variation between these two limits is linear:

$$\Delta\sigma = E\frac{d}{R} \quad \text{or} \quad \Delta\varepsilon = \frac{d}{R} \qquad (2.47)$$

where d is the thickness of the film under consideration and R the radius of curvature.

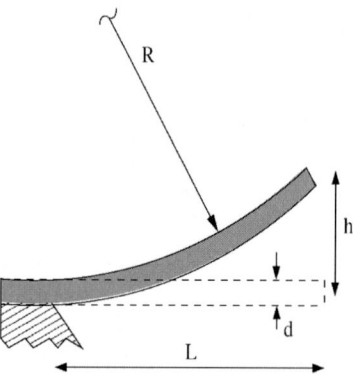

Figure 2.18: Relaxed cantilever beam under strain gradient.

As demonstrated in p. 62, radius of curvature R can be extracted from Eq. 2.31 where $L/2$ in case of the curvature of a total wafer is replaced by L for a deflected cantilever and $h_{\max}$ replaced by h (see Fig. 2.18),

$$R = \frac{L^2}{2h} \qquad (2.48)$$

From Eq. 2.47 and 2.48 we can express the stress gradient $\Delta\sigma$ per unit of thickness d as

$$\frac{\Delta\sigma}{d} = E\frac{2h}{L^2} \qquad (2.49)$$

where $\frac{\Delta\sigma}{d}$ is expressed in Pa/μm or MPa/μm [84]. A cantilever beam exhibiting a concave shape, i.e. a bending away from the substrate, is associated with a positive stress gradient. A negative stress gradient is associated with cantilever beams bending towards the substrate, i.e. a convex beam [57].

In comparison with the substrate curvature method (Stoney equation) where the stress is integrated over the whole thickness of the film, the cantilever equation only gives the

2.3. STRAIN MEASUREMENTS USING MICROMACHINED STRUCTURES 89

difference of strain between the top and the bottom of the thin film. In both cases, stress is transformed to a curvature when the film is relaxed. Thanks to some simplifications, Stoney equation gives the stress without the knowledge of the biaxial modulus but it remains useful for extracting the average strain. On the contrary, in case of cantilevers, the strain gradient is first calculated and the Young's modulus of the film remains useful for extracting the stress gradient.

The beam curves in a circle of radius R, a long beam will describe a longer part than a shorter one, and a high stress gradient will curve the cantilever with a lower radius than a lower stress gradient. One length of cantilever will cover a wide range of stress gradients but if its deflection becomes higher than its radius R, it will not be measurable anymore. So, an array of cantilevers is needed to allow the measurement of a large range of stress gradient. A high stress gradient needs for short beam to be measured. On the other hand, long beams will be useful to measure lower stress gradient.

Cantilevers were processed similarly to the previous microstructures taking into account the undercutting properties of silicon by TMAH at convex corners. All beams in the array have been carried out in four widths: 2, 5, 10 and 15 μm. Theoretically, the beam width has no effect on the stress gradient but in practice, it is not completely the case as shown for clamped-clamped beams. Wider beams were showing a slightly lower deflection than narrow beams due to their higher stiffness probably. It is not so critical in our case where only relative measurements are interesting for the comparison of the strain gradient between two locations or between two processes. Therefore, measurements will always be performed on same widths.

Deflections h were measured using optical microscope comparing the focusing on the two tips of the cantilever. Deflections of 12 μm were measured for 100 μm length and 10 μm wide cantilevers made in thermal oxide, giving a radius of curvature R equal to 416 μm. A strain gradient of 0.24 %/μm, i.e. a stress gradient of 170 MPa/μm for thermal oxide was calculated therefrom. For a thickness lower than 1 μm (0.4 μm), it is more clear to express this value as 17 MPa/0.1 μm, i.e. a difference of stress between the top and the bottom of the layer of 68 MPa.

In comparison with the average stress measured in thermal oxide (-331 MPa) over the total thickness, the 68 MPa represent only the variation of stress between the top fiber and the bottom fiber of the layer. The previous measurement methods gave only an average stress in thermal oxide, the stress on the neutral fiber at the middle of the layer as shown in Fig. 2.17 (before release). When a cantilever is released, the average stress is

Figure 2.19: SEM view of an array of 10 μm width cantilevers builded in thermal oxide on etched silicon cavity. Longest showed beam is 150 μm long.

removed because the beam is free to elongate or shrink without introducing any visible deformation. A cantilever therefore only measures the stress gradient which is exhibited when the cantilever curves to its relaxed position. So, the total stress in thermal oxide can be expressed as -331 ±34 MPa.

Measurements were also performed with interferometry giving the profile shown in Fig. 2.20. Measurements were possible only for short beams (50 and 70 μm) featuring low deflections due to the lack of field depth in case of too high out-of-plane displacements. To increase the depth of field, a lower magnification must be chosen leading to a lack of lateral precision and contrast. But in extrapolating the two results we have obtained by quadratic interpolation, we were able to extract the radius of curvature of the beams. By this way, our last result of 12 μm of deflection for 100 μm long beam measured previously by optical microscopy was confirmed.

In other way, we observed as shown in Fig. 2.21 a large discontinuity at the beginning of the beam near the anchors. This is physically inconsistent but can be explained as followed: the light is reflected on the silicon side of the cavity instead of being reflected on the transparent beam. This phenomenon occurs only close to the anchors where the distance between the beam and the silicon layer is shorter. To calculate the profile of the beam and the maximum deflection, we therefore must take it into account summing the minimum and the maximum deflections of the beam. We observed the same behavior for clamped-clamped beams and calculations were performed in the same way.

2.3. STRAIN MEASUREMENTS USING MICROMACHINED STRUCTURES 91

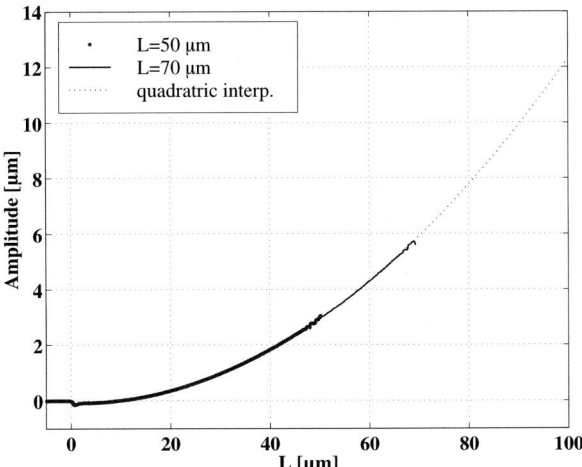

Figure 2.20: Thermal oxide 5 μm width cantilever profile for two lengths and quadratic interpolation showing the curvature of the beam.

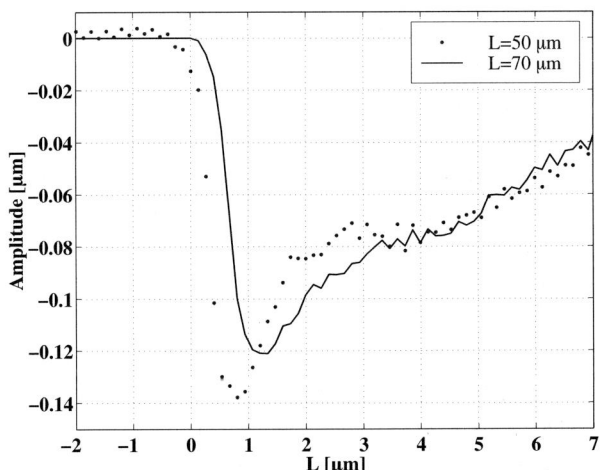

Figure 2.21: Detailed view of the discontinuities appearing between the anchor and the beginning of the beam.

Our calculations of the stress gradient from cantilevers structures are in fact a mathematically calculated average value. This value results in a similar deflection curve to that of the actual stress profile. The stress profile throughout the sample can be very complicated and depends a lot on the growth. Thus, a change in stress in a very small section of the layer can have an important effect on the deflection and, correspondingly, on the determined measured stress gradient [57]. We indeed supposed a linear stress profile between the top and the bottom of the oxide film since our calculations only give the difference of stress from the top to the bottom of the film. For the physical explanation of the stress behavior inside the layer, it could be therefore attractive to determine a complete stress profile throughout the thickness to show that profile is not so simple. As stress gradients become particularly acute as the film becomes thinner [77], an idea consists in measuring the stress gradient in a film at different thicknesses by progressive thinning [84][57]. The stress profile through the thickness can be calculated by this way from the values of the stress gradient. Afterwards, the average stress of the measured profile can be confirmed as the total stress in the thin layer. But due to the limited vertical resolution of this technique, this method is increasingly inaccurate for thinner layers [57].

We generally assume that average stress has no effect on the bending of a cantilever. Fang and Wickert [77] demonstrate the opposite. The portion of the beam clamped to the substrate (the anchored end) undergoes some in-plane deformation (expansion or contraction) under an average stress. The deformation of this portion of the film bonded to the substrate leads to a slight rotation of the released cantilever at the anchored tip as modeled in [77]. This phenomenon is exploited to find the total residual stress. So, under a combination of average stress and stress gradient as in thin oxide layers, a released cantilever will deflect out-of-plane, with its far field curvature being exclusively generated by the stress gradient (as measured above) and with an initial slope determined by both the average stress and the stress gradient. Characterization of the measured deflections by interferometric profilometry in this manner allows to estimate the total residual stress using the model of [77]. Only one single cantilever instead of a complete array is needed in this case to measure simultaneously average stress and stress gradient. But as observed in [57], the measurement of the slope of the beam at the suspension is very complicated and time-consuming and on the other hand, the effects of the rotation introduce very small error in the stress gradient value. Moreover in our case, the discontinuities occurring under interferometric measurements at the anchored tip of the beam do not allow to confirm this slope. For these reasons we did not consider it in our analysis.

This previous method can be extrapolated to improve the accuracy of the bilayer

2.3. STRAIN MEASUREMENTS USING MICROMACHINED STRUCTURES 93

method [68]. The bilayer method consists in depositing the thin film layer of interest on a cantilever beam used as a base material so as to form a bilayer sandwich-like structure. The stress level in the film layer is then deduced by measuring the deflection of the bilayer cantilever taking into account the inital curvature of the base cantilever measured prior to the film layer deposition. This technique is especially useful to characterize very thin films that are not stiff enough to be used as cantilever beams. Nevertheless, as the thin film must be deposited after the release of the base layer, this technique is only usable for metallic sputtered films.

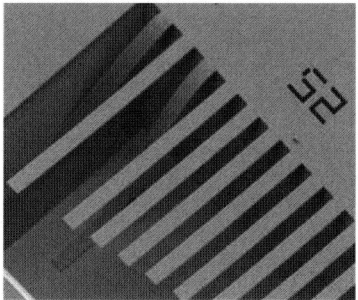

Figure 2.22: Stiction appearing between nitride cantilever and silicon on the bottom of the cavity after water rinsing and drying in methanol.

Finally, it appears interesting to us to introduce the main problem appearing in the fabrication of this kind of suspended cantilever beams which is know as the **stiction**, i.e. the permanent adhesion of the beam to the substrate if their surfaces come into contact. The phenomenon is usually divided in two stages: first, a mechanical collapse mainly caused by the attractive capillary forces, and second, a permanent adhesion to the silicon substrate caused by the hydrogen bonding [85]. Hydrophilic silicon has a large number of surface -OH groups so that hydrogen attraction occurs between silicon and suspended cantilever when fluid as water, nitrogen or oxygen is trapped in the gap, causing their sticking [11][86]. The use of methanol, which has a lower surface tension than water, can be applied after the final water rinse to displace the water [86]. It reduces the stiction but does not avoid it completely as depicted in Fig. 2.22 showing two nitride cantilevers sticking on the bottom of the silicon cavity after methanol rinsing. The best way to suppress this effect is the use of critical point drying to dry the samples after their rinsing. This method takes advantage of the fact that, under the correct conditions of temperature and pressure (the "supercritical region"), the liquid and vapor phases cease existing as distinct states. When this occurs, the interface between them is eliminated and, after transitioning directly to the gas phase, the gas can been gently vented without disturbing the structures [11]. In this process, the wafers immersed in methanol are brought into a pressure vessel at room temperature. Liquid CO_2 is then used to replace methanol during a rinse cycle. The temperature of the liquid CO_2 is then raised above its critical point and the pressure vessel is finally vented and the CO_2 gas escapes. A liquid to solid interface is never formed during the process, and hence surface tension is completely suppressed [86].

Multilayer cantilever beams

Stress gradient can be intrinsic as shown above but can also appear when two or more layers (with uniform or non-uniform stress) are stacked together due to the difference between the average residual stress of each film in the multi-layer. In this case, stress gradient is mainly caused by a gradient in the thermal expansion coefficient of the stacking layers. So the magnitude and sign of the resulting stress gradient can be different for each combination of films [65]. In this case, the stress gradient is more difficult to extract since we need to know the Young's modulus of the stacking to calculate the stress gradient from the strain gradient. Nevertheless, the strain gradient can be calculated using the following equation derived from Eq. 2.49:

$$\frac{\Delta \varepsilon}{d} = \frac{2h}{L^2} \tag{2.50}$$

where d is in this case the total thickness of our multi-layer.

An average Young's modulus of the stacking is difficult to extract. The stacking is not an alloy and it would be wrong to calculate an average Young's modulus (of the combination) as the linear interpolation between each constitutive layer (as calculated in [84] for SiGe layers for instance). In [65], the Young's modulus of stacked films is measured on composite cantilevers by nano-indentation. It is presented as a technique to determine the Young's modulus of each thin film constituting the composite stack, without accessing the individual layers during processing. To perform that, the summation formula $E_{av}I_{av} = \sum_{j=1}^{n} E_j I_j$ is used were E_{av} and I_{av} are respectively the Young's modulus and the moment of inertia of the stacking, and E_j and I_j, the individual Young's modulus and the moment of inertia of the jth film. A different formula to calculate the average Young's modulus is presented in [79] as $E_{av}A_{tot} = \sum_{j=1}^{n} E_j A_j$ where A_j and A_{tot} are respectively the area of the jth material in the cross section and the total cross sectional area.

2.3. STRAIN MEASUREMENTS USING MICROMACHINED STRUCTURES 95

The calculation of an average Young's modulus is a solution more or less accurate to extract the stress gradient from the strain gradient. A more rigorous solution would be to measure the complete stress profile through the thickness of the combination by progressive thinning as exposed above. It is nevertheless very complicated to implement and not really necessary in our case. The direction of the cantilever deflection, its measurement and the calculation of the strain gradient is enough for us.

(a) (b)

Figure 2.23: SEM view of an array of 10 μm width cantilevers builded in ON (a) and ONO (b) stacking.

Strain gradient measurements were performed on O-N and O-N-O sandwich structures. Both combinations showed positive stress gradient, or a deflection away from the substrate as shown in Fig. 2.23. The deflections were measured by focal plane adjustments on optical microscope but could not be confirmed by interferometry as explained above. As observed when measuring gradient in oxide layer, wider beams showed always a lower deflection than narrow beams probably due to their higher stiffness. Measurements were performed on same width (10 μm) for more relevant comparisons. A radius of curvature going from 94 to 104 μm was measured for O-N cantilever beams having a deflection smaller than their radius of curvature (before the 25th as shown on Fig. 2.23(b)), i.e. a strain gradient from 0.96 to 1.06 %/μm. In the same way, we measured a radius of curvature from 128 to 130 μm for O-N-O sandwiches leading to a strain gradient from 0.77 to 0.78 %/μm.

As shown in Fig. 2.23(b), PECVD oxide on top of the cantilever was overetched sideways during the etching of the thermal oxide due to the lower density and the higher etch rate of PECVD oxide versus thermal oxide. This problem can not be avoided but is not critical since the outcoming geometry does not affect the measurement of the strain gradient. It is clearly visible also that the combination O-N-O has a lower strain gradient than the same combination without PECVD oxide (O-N). It confirms that densified PECVD oxide added on the top of the structure leads to a compensation of the average stress and the stress gradient in the structure.

Table 2.8 summarizes the results of our strain and stress gradients measurements. We were able to extract the stress gradient from the thermal oxide only thanks to the knowledge of its Young's modulus. On the other hand, only strain gradient could be measured for the other layers and layers combinations. LPCVD nitride and PECVD oxide did not show any significant strain gradient.

Table 2.8: Approximated values of strain and stress in mono and multi-layers for a given cantilever width.

Layer	strain gradient [%/μm]	stress gradient [MPa/μm]	curvature
Thermal oxide	0.24	170	+
LPCVD nitride	negligible	negligible	/
PECVD oxide	negligible	negligible	/
Sandwich O-N	1.01	/	+
Sandwich O-N-O	0.78	/	+

Discussion on multilayer membranes

The strain gradient measured in O-N-O combination is the same strain gradient as the one expected in our membranes. In static equilibrium, it will not affect the robustness of the membrane, it is just interesting to know that the top is in average more tensile than the bottom as demonstrated by the cantilever. If the membrane is subjected to a load from the top or from the bottom, things will be different. The membrane will tend to buckle up or down under the force or the pressure in spite of the fact that the membrane is tensed by an average internal tensile stress (of +80 MPa). If the membrane curls up (Fig. 2.24), the top layer will be more elongated than the bottom layer (due to a higher radius of curvature) and therefore subject to a more compressive stress than the bottom layer. Forced buckling in this case will compensate the positive stress gradient present in the stacking. On the contrary, if the membrane is curling down, the opposite will occur, resulting in an additive compressive stress at the bottom. In this case, cracks could appear in oxyde layer if deflection is too high but could be stopped by the nitride layer at the

2.3. STRAIN MEASUREMENTS USING MICROMACHINED STRUCTURES

middle of the stacking thanks to its high tensile stress. In the case of our application, the membrane will curl up due to the displacement caused by the heating microheater (see chapter 1 of part III) and could therefore be expected to be more robust in our case than a unilayer membrane.

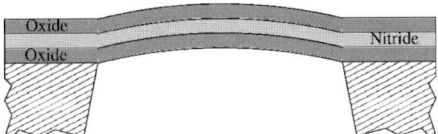

Figure 2.24: Membrane under deflection.

To finish, it can be interesting to note here that a pressure applied on a micromachined membrane can be used to evaluate directly its average stress. A differential pressure applied across the diaphragm causes its bowing outward. The vertical displacement at its center is then measured using an optical interferometer and is related to the stress [1]. This method could be attractive in our case to directly extract the average stress in our performed membrane (at the end of a gas sensor process for instance) but it is complicated to perform and needs to know the Young's modulus and the Poisson ratio of the stacking to extract the final average stress. Moreover, calculations derived from equation shown in [1] are complicated and need to find geometry constants. The previous studied methods will therefore remain the best combination even to extract the stress in a membrane.

2.3.4 Summary and outlook

Clamped-clamped beams and ring-and-beam arrays gave accurate results requiring only interferometry to perform out-of-plane deflection measurements. On the other hand, microgauge measurements need SEM imagery to measure the displacement of the tip of the indicator beam reported to the scale. Both methods were confirmed easy to perform. The accuracy of the first method is limited only by the precision of the layer thickness measurement. In the second method, the measurement is independent on the layer thickness but it can be less accurate when the gauge and its scales are overetched, as shown in Fig. 2.16(a).

Using clamped-clamped beams in post-bukling state tends to slightly underestimate the actual compressive stress value probably due to the oblique geometry of the beams as explained above. However measured values on thermal oxide are reproducible with a narrow dispersion as expected for a high quality thermal oxide. Ring-and-beam structures showed a wide variation of stress value across the wafer due to the lower quality of the deposited LPCVD nitride film and the limitation of the method accuracy. The overestimations observed for oxide gauges in comparison with results obtained by previous methods are probably due to patterning problems occurring during HF etching and to the high stress gradient present in thermal oxide. Again results are consistent and relatively constant on the whole wafer as shown previously with beams. Results in LPCVD nitride agree reasonably well with the range measured using rings-and-beam structures thanks to the higher resolution of the nitride plasma etching and its lack of stress gradient. Finally, cantilevers showed a significant intrinsic stress gradient present in our thermal oxide monolayer and how the stacking of multilayers with different average stress can affect the final stress gradient of a beam or a membrane.

Realizations of dielectric clamped-clamped beams or cantilevers using sacrificial etching of silicon can be found in literature [79][65][27][78]. However, most of the time these structures are based on isotropic echting of silicon. In this work, complete arrays of dielectric clamped-clamped beams, ring-and-beams, microgauges and cantilevers were released for the first time using anisotropic TMAH etching instead of isotropic etchants. TMAH etching requires a wise design to take into account its undercutting properties but gives at the end a better control of the anchored tips without a critical control of the etching time. Microstructures are therefore easier to process while giving more reproducible results.

Our microstructures array was also used to characterize stress in pure evaporated **aluminum**. It is out of the scope of this chapter but is interesting to note for the following reason. Aluminum microstructures can not be built on sacrificial oxide layer due to the very bad selectivity of HF versus aluminum. It is the same for most isotropic and anisotropic silicon etchants making difficult the use of silicon as sacrificial layer. In our case, thanks to our improved TMAH solution giving a high selectivity versus aluminum, we can use the previous microstructures to characterize this material. Microgauges measurements were performed on 900 nm thick evaporated aluminum layer. Measurements yielded values varying around zero from **-80 to +70 MPa**, therefore either compressive or tensile. These values were confirmed by wafer curvature measurements yielding values between -30 and +60 MPa. Regarding results of the literature, we find in [27] that aluminum has a low melting point and a corresponding high diffusion rate even at room temperature and is usually fairly stress free. Our stress values varying around zero seem therefore consistent. However, in comparison with measured values for dielectric

2.3. STRAIN MEASUREMENTS USING MICROMACHINED STRUCTURES 99

layers, we observed a wide dispersion across the wafer and a non reproducibility from one process to another. An explanation of these dispersions would be the high solubility of the silicon into the pure aluminum film affecting its properties and its homogeneity. It could be better therefore to deposit a very thin oxide layer on silicon before the aluminum evaporation which could be removed at the end without damaging aluminum to perform the measurements. Another solution would be to deposit an aluminum doped with 2 % of silicon by sputtering as used in standard IC processes. But sputtered metals can be expected to be more stressed as confirmed in [59].

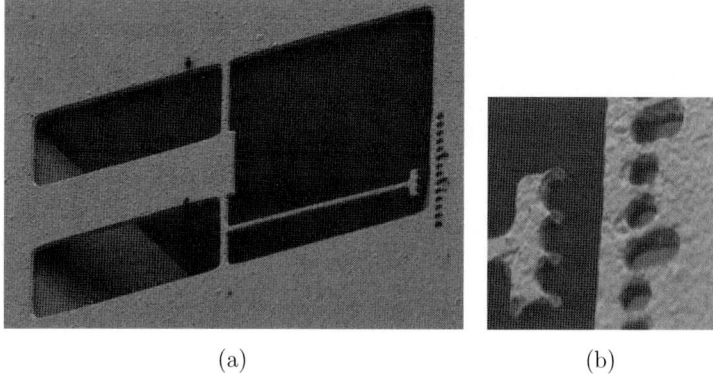

(a) (b)

Figure 2.25: Stress measurements in aluminum layer on silicon as sacrificial layer (a) microgauge in aluminum and (b) detail of the tip of the indicator beam.

It would be possible to characterize nitride using oxide as sacrificial layer thanks to the great selectivity of the oxide etchant (HF) versus nitride. The experiment was not performed for a thin tensile nitride layer but for a compressive thicker **polysilicon** layer for which we have no other choice. We focused especially on microgauges [55] and clamped-clamped beams as depicted in Fig. 2.26. A 2.5 μm thick polysilicon layer deposited at 625°C under 90 sccm of SiH_4 was investigated. Deflection to the right of 4.72 μm was measured on the microgauge and buckling was detected on the clamped-clamped beam area between lengths of 100 and 150 μm (Fig. 2.26). Both methods gave finally the same strain value around 0.13 % or a compressive stress value of **-210 MPa** using a Young's modulus of 160 GPa. This value is difficult to compare with literature due to the wide variety of polysilicon layers but is in good agreement with measurements performed by wafer curvature method over several processes (around between -180 and -220 MPa).

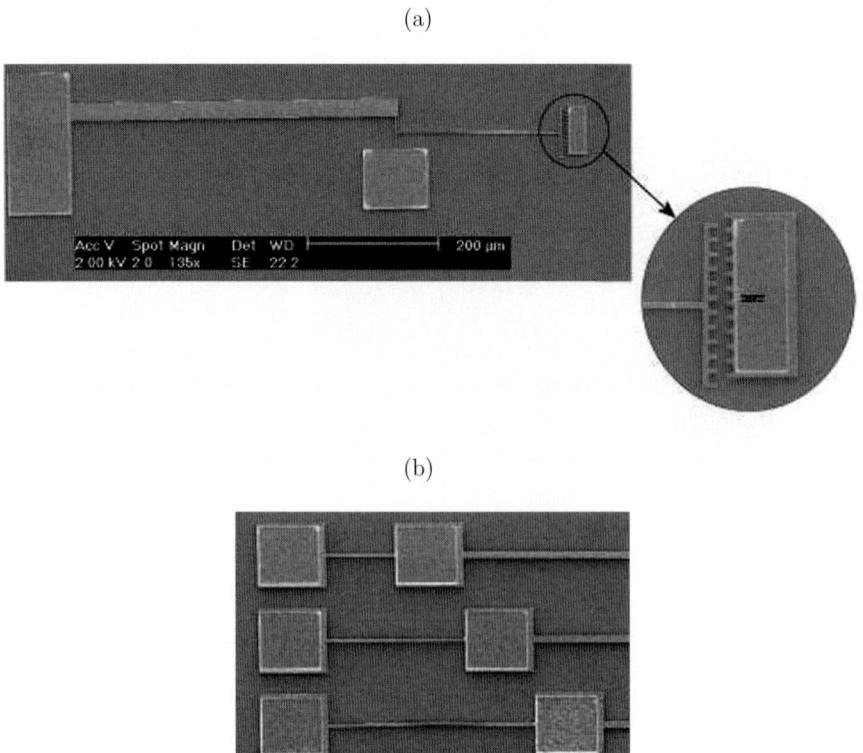

Figure 2.26: Stress measurements in polysilicon layer using oxide as sacrificial layer (a) microgauge with $L_{tb} = 500$ μm, $w_{tb} = 30$ μm, $L_{sb} = 50$ μm, $w_{sb} = 3$ μm and $L_{ib} = 250$ μm, $w_{ib} = 4$ μm; (b) part of the clamped-clamped beam array (width 3 μm) around the critical beam, lengths are respectively from the first to the last 50, 100 and 150 μm. The buckling is clearly occuring between the second and the third beam.

2.3. STRAIN MEASUREMENTS USING MICROMACHINED STRUCTURES

In comparison with microstructures fabricated on silicon, in this case, a sacrificial layer must be deposited and patterned using an additional mask. A 2 μm thick PECVD oxide layer was deposited as sacrificial layer, and patterned with HF to define the anchors of the future microstructures. This step is critical and depends strongly on the density of the used oxide. A weakly densified oxide will be easier to release at the end than a dense oxide but will be more damaged sideways during the etching of the 2 μm thick anchors. A densified PECVD oxide having an etch rate around 100 nm/min seems a good compromise for this purpose. After this etching, polysilicon was deposited and patterned by plasma. The release of the microgauges and clamped-clamped beams was performed in wet HF, rinsed in water and finally dried in methanol. Stiction was a critical problem in this case due to thin gap between the suspended microstructure and the substrate. Methanol is one solution but vapor HF releasing [87] or critical point drying would be more effective.

Microstructures using oxide as sacrificial layer are the unique solution to measure stress in polysilicon since polysilicon layer would be completely etched during a potential silicon release. But if we have the choice as for nitride for instance, the use of silicon as sacrificial layer remains more powerful. Silicon release is more economical in time and in materials, it avoids stiction problems, and the critical patterning of anchors is by-passed. Another drawback of the use of oxide as sacrificial layer is due to the thin gap between the suspended beams and the substrate. Stress in very thin films is more difficult to measure since they buckle more often out of the plane and are for this reason more attracted to stick on the substrate when buckling downward. It is therefore needed to characterize films thicker than the film of interest. The measurement will remain valid in the absence of stress gradient but if a gradient occurs, the measured average stress will be not so consistent. It is the case in polysilicon layer as demonstrated in the literature [27].

To conclude, the new techniques presented here to extract the stress using microstructures present a lot of advantages as compared to most realizations proposed in literature:

- Anisotropic etching avoids geometric problems at the anchors and tedious control of the etching time,

- Silicon used as sacrificial layer is more economical in terms of time and cost since it uses only one photolithographic step and only one mask, it also allows the ability to measure stress in layers as silicon oxide,

- The formation of deep cavity under the microstructures avoids stiction problems,

- The high selectivity of our TMAH based solution allows to extract the stress in aluminum layers,

- The characterization of the film in its actual thickness gives more accuracy, especially when stress gradient is present.

2.4 Final conclusions

An in-depth study of the stress and stress gradient in thin dielectric films was performed and explained along this chapter. After a theoretical recall of the mechanics involved in thin films, two well known methods were demonstrated and developed by experiments from a new approach. Both methods allow for stress measurements on dielectrics layers using the same thin dimensions as in target applications and for this reason, a small area was needed to integrate microstructures arrays. Stress and strain gradient in multilayers were also measured to be able to finally extract the average stress and strain gradient in a stacked dielectric membrane to predict its robustness. Silicon was chosen as sacrificial layer allowing the extraction of the stress in oxides and microstructures release was performed anisotropically instead of isotropically for a lot of detailed advantages.

Two methods were shown complementary: microstructures allow microscopic and local stress measurements and can be integrated in situ in any processed wafer, while curvature method gives average macroscopic results and needs additional test wafers to be added to the processed ones. Table 2.9 summarizes our results, demonstrating that both methods gave relatively similar results thanks to the homogeneity and the high quality of the films of interest. The overestimation of the stress measured using oxide gauge and the under estimation of the clamped-clamped oxide beams were explained by the high stress gradient present in the film.

Table 2.9: Summary of our stress measurements results.

Layer	Average stress [MPa]			Strain gradient [%/μm]
	Stoney	Bridge/Ring	Gauge	
Thermal oxide	-331	-300...-330	-460...-490	0.24
LPCVD nitride	860	840...1015	795...950	0
Dens. PECVD oxide	-172	-160...-190	x	0
Sandwich O-N	174	x	x	1.01
Sandwich O-N-O	80	x	x	0.78

2.4. FINAL CONCLUSIONS

Both methods can therefore be compared and used according to the test purpose. Curvature measurements will calculate a value of the average stress without needing knowledge of the elastic constants of the film under interest, while microstructures will extract the average strain in the film. The first method also allows to measure exclusively the average stress in multilayers, while the second allows to extract the actual strain gradient value in these multilayers.

In our case, average stresses were calculated from the measured strain values using well established elastic constant values. But these constant values can vary from one piece of equipment to an other as well as between two processes. Difference in stress values extracted from wafer curvature measurements and microstructures could therefore be explained from a non accurate knowledge of materials properties. However, the strain of a layer having unknown elastic constants could be advantageously extracted from microstructures and the stress, from curvature method, so that finally, the Young modulus and Poisson ratio of this layer can be calculated. The complementary of the two methods is therefore fully demonstrated.

We also measured an average stress of -363 MPa in the buried oxide of a Unibond SOI wafer. We demonstrated by this measurement that this layer can advantageously replace the first thermal oxide layer of our membrane for gas sensors applications. Finally, one wafer was used to follow all steps of a gas sensor process to finally measure by wafer curvature technique, an approximation of the expected stress in the future membrane of our gas sensor. For more accuracy in next processes, we could put arrays of microstructures on the wafer to extract the final average stress and the strain gradient in each layer constituting our multilayered membrane.

Part III

Microsensors

Chapter 1

Low power microhotplate as basic cell

1.1 Introduction

In the recent years, thin-film microhotplates have been emerging as a topic of considerable interest for an extremely wide range of applications such as gas sensors [73][76][88][19]. For such structures, a microheater based on Joule heating is needed to uniformly heat a square or circular area as it will be later explained. The major challenges involved in the design of these microhotplates are the needs for high thermal uniformity to increase the sensitivity and the selectivity of the heated layers to specific gases, low-power consumption for portable applications, high mechanical and thermal strength to increase robustness in harsh environments and especially compatibility with standard IC processes to enable co-integrated microsystems and to allow low cost batch production. Finally, the ability to reach temperatures up to 700°C is an asset to perform the on-chip annealing of the deposited gas sensitive layers after packaging. All of these requirements are detailed in the next sections.

Numerous research groups have mentioned a great interest for thin dielectric membranes in term of power consumption for such microhotplates [73][76][88][19]. The power consumption of a 1 μm thick SiO_2 membrane was reported to be 75 % less than that of a 1 μm thick silicon membrane for the same maximum temperature [73]. Dielectric membranes exhibit lower thermal conductivities than silicon but need to be carefully designed to decrease the high residual stresses of their constitutive layers. In Chapter 2 of the first part, we designed and fabricated a 1 μm thick stacked dielectric membrane (SiO_2 - Si_3N_4 - SiO_2) showing very low residual stress (tensile stress of 80 MPa) and low strain gradient

(0.78 %/μm). Its **high strength** was expected and will be demonstrated in this chapter when covered with a polysilicon microheater. Its lower thickness in comparison with most recent publications (see Table 1.3) allows to reach lower power consumption as it will be demonstrated.

To reduce cost and process complexity, our design offers the challenge to be **fully compatible with SOI-CMOS processes** with minor changes. Many micro-hotplates (and gas sensors) reported in literature were referred to as "CMOS compatible" but most of them used layers which are not inherent in a standard IC fabrication (such as Platinum used for the heater, gold for the electrodes) or used technological steps which are too complicated (and therefore too expensive) to be co-integrated with CMOS circuits (such as KOH etching, silicon spreader underneath the membrane). Furthermore, only a few have been actually produced in standard CMOS process, based on the layers available in such technology. Nevertheless, each of them suffers of one or more critical disadvantages: [89] and [90] reported high power consumption (47 mW at 400°C with an efficiency of about 8°C/mW) and a maximum operating temperature of 400°C due to an aluminum heater; [91] revealed the same problems; [92] and [93] demonstrated the monolithical co-integration of the sensor with its interface circuitry but used chromium and gold interdigitated electrodes to be able to increase the operating temperature and showed a thermal efficiency of only 10°C/mW; [94] reported an other design co-integrated with a CMOS drive circuitry, consuming only 9 mW to heat a 80x80 μm^2 area at 400°C but featuring Pt/Ti films employed as both heater and temperature sensor; [95] reported an interesting and new design which implements an on-chip temperature controller to regulate the temperature of the membrane and which consumes only 4.8°C/mW to heat a 300x300 μm^2 area, but provides a silicon island underneath the membrane etched with KOH; finally, a co-integration was tried in [96] in two steps: the CMOS circuits were structured on one half of the wafer, while the second half was protected. The second part was then post-processed with the gas sensors, while the CMOS were simply protected with a photoresist layer. It is not a real co-integration in this case since the two parts were manufactured simultaneously.

It will be demonstrated that only two more masks are needed to fabricate in same time our microheater on its membrane and integrated circuits in SOI-CMOS technology. Furthermore, it will be shown how easier and more suitable it is to co-integrate the sensor and its electronics in SOI-CMOS technology instead of another process. To allow the full compatibility between both technologies, the membrane will be released using TMAH-based (as Tetramethyl Ammonium Hydroxide) etching technique, much praised

1.1. INTRODUCTION

in Chapter 1 of part II. We will finally demonstrate in Chapter 4 of this part, that the additive steps do not affect the physical and electrical behaviors of the CMOS circuits. We will report for the first time a detailed study concerning the impact of each gas sensor post-processing step on CMOS circuits.

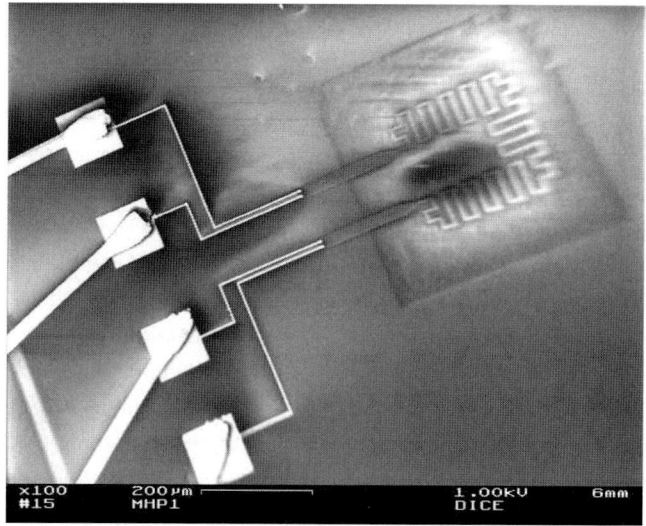

Figure 1.1: Packaged polysilicon microheater on 400x400 μm^2 membrane. On chip aluminum connections (bright lines) and gold wires to the package can be seen.

Along this chapter, the design, fabrication and experimental results will be presented. In first part, the motivations of our realization will be described in details, especially the fabrication of smart gas sensors. A description of base materials and a detailed study of our design using ANSYS numerical electro-thermal simulations will be presented. The microheater fabrication is then described on classical silicon substrate and compared with its fabrication using SOI (Silicon-on-Insulator) technology to explain its co-integration. Finally, numerous experimental results of our microheater characterizations are thoroughly discussed.

1.2 Motivations

1.2.1 Gas sensors

As mentioned in the introduction, our microhotplate is developed and optimized to eventually build a fully integrated smart gas sensor in SOI-CMOS technology. The principle of this kind of sensor lies in a sensitive layer (typically a metal-oxide layer such as SnO_2 or WO_3)for which the resistivity is modulated at a working temperature between 200 and 400°C in presence of gases such as NO_x, SO_2 or CO due to adsorption reactions on its surface. The maximum sensitivity is achieved for each material at a specific temperature which depends on the gas to be detected [97].

For such sensor, a microhotplate is necessary to heat the sensitive layer and a temperature sensor is needed to control the temperature during operation. Electronic circuitry such as an operational amplifier must also be added nearby to the sensor to monitor the resistivity modulation as well as to perform in-situ control and monitoring. SOI technology is uniquely suited for micropower as well as high temperature and radiation circuit performances [6]. SOI substrates also offer a lot of advantages towards the joint fabrication of the membrane with the close electronics. The typical 400 nm buried thermal oxide of the SOI wafer can constitute the first layer of the membrane and also the natural etch-stop layer for the TMAH backside silicon etching, while the upper monocristalline silicon film can integrate the electronics close to, but totally isolated from the microheater. It is therefore more suitable to integrate this microsystem using SOI-CMOS technology combining process easiness and high circuit performance.

To complete a gas sensor on top of a microheater, interdigitated sensing metallic electrodes must be patterned on the active area of the membrane. Thick SnO_2 or WO_3sensing film must next be deposited on the released membrane by sputtering or drop coating technique. Finally, annealing of the active layer must be performed at **high temperature** (700°C) to increase their stability and performance [98]. The membrane therefore needs to be very robust to allow metal-oxide deposition and to be heated up to a very high temperature during 15 minutes. As this heating can not be performed on the whole wafer due to its non-compatibility with CMOS circuits, annealing "on-chip" of the gas sensitive material is proposed after sensors packaging. This technique has already been proposed in [99] but using a Platinum microheater resistor.

For gas sensors, **high thermal uniformity** is also critical to increase sensitivity and selectivity to gases. As reported in [76], the temperature gradient over the sensitive area

1.3. MATERIALS SELECTION

should not exceed 50°C to conserve good sensitivity performance. [88] recommends a temperature gradient lower than 25°C. The originality of our design also results from its polysilicon loop shaped heater (Fig. 1.1) designed to reach a fair thermal uniformity without requiring the addition of a silicon spreader underneath the membrane. This latter technique was chosen in most realizations [88][19] but strongly increases the power consumption, the microheater inertia and the process complexity.

As demonstrated in [100], **low power consumption** and **small size** are finally critical not only to allow portable applications but also to integrate multiple sensing elements on a single device for implementing sensor arrays. Semiconductor gas sensors lack of a good selectivity and usually an array of sensors on the same chip maintained at different temperatures is necessary for obtaining data that, after being properly processed, allow a good discrimination between a target gas and other interfering gases [19][97]. Note that the selectivity to gases as well as the low power consumption can also be improved by the modulation of the microhotplate operating temperature [101]. When the operating temperature is pulsed, different gases give different dynamic responses (i.e. signatures) depending on the used temperature cycles [90].

1.2.2 Other sensors

Microheaters can also be used in many applications such as e.g. flow sensors as will be studied in the next chapter. Coupled with bulk micromachining and wafer bonding, it can heat liquid fluids. Polysilicon when heated at high temperature can emit light in visible as well as in infrared and can be exploited to be used as IR source as proposed in [102]. This behavior was partially observed in our case in visible spectral and will be reported later. However, this device could not be used as IR sensor due to a too low sensitivity of polysilicon. Finally, microheater could be used as pressure [103] and ultrasonic sensors [104] in combination with e.g. piezoelectric resistive transducers.

1.3 Materials selection

1.3.1 Membrane

The materials chosen for the membrane of the microhotplate should combine low thermal conductivity (i.e. small thickness) with high mechanical strength (i.e. large thickness) [19]. While insuring compatibility with a SOI-CMOS process of interest for micropower or high-temperature applications, the 400 nm thick buried thermal oxide of the SOI substrate can advantageously constitute the first part of the membrane. A second part of the membrane stack can be constituted by the densified PECVD oxide layer related to the interconnect dielectric between the polysilicon and aluminum layers of the CMOS process. For mechanical robustness, a tensile LPCVD nitride layer needs to be added between the two oxide layers and its thickness must be calculated to carefully compensate the high compressive stresses of the oxide films.

A detailed study about the stresses in our dielectric membrane was presented in Chapter 2 of the first part. We optimized the film thicknesses at 300 nm for the LPCVD nitride layer and 300 nm for the densified PECVD oxide layer to reach a total membrane thickness of around 1 μm (with the 400 nm thick buried oxide layer). We concluded that this kind of membranes shows a slight tensile residual stress around 80 MPa and a small strain gradient of 0.78 %, just enough to obtain a robust and flat membrane. Our simulations will confirm in next section that the chosen thickness offers a good trade-off between membrane robustness and power-consumption.

1.3.2 Microheater

The design chosen for the heater is a loop-shaped phosphorous doped polysilicon resistor as optimized in next section by finite element method (ANSYS software) to achieve low power consumption and high thermal uniformity. The 340 nm thick phosphorous doped polysilicon commonly used for the gate in CMOS and SOI-CMOS fabrication was used as base material for the heater.

The doping level was imposed by the IC process and yields a sheet resistance around 27 $\Omega/\square$. At this doping level, polysilicon shows a positive temperature coefficient of resistance (TCR) which, as well as Platinum or other conductors, unfortunately leads to hot spots on the heater [100]. If the doping was lower, it would have semiconductor behavior and a negative TCR, which would lead to a more uniform temperature distribution over the heater but a higher power consumption. Nevertheless, it will be demonstrated that the TCR of our polysilicon layer is lower than pure conductors such as platinum and is by the way a good compromise for microheaters. In comparison with platinum often used as heater [99][105], polysilicon also features the advantages of a lower cost and of being fully compatible with CMOS technology. Similar conclusions were reported in [106] and [107] which compared the heating properties of polysilicon and platinum microheaters.

1.4. THERMAL DESIGN

The main drawback of polysilicon is the change of its resitivity (and TCR) at high temperature [108][109] as well as its poor long-term thermal stability [105][110][97][111]. It is therefore crucial to characterize and understand its behavior at high temperatures as well as versus time in order to be able to foresee its evolution.

Finally, the stress in our doped polysilicon was measured compressive (around -200 MPa) using micro-gauges and wafer curvature technique as reported in Chapter 2 of the first part. So high compressive stress could damage the mechanical robustness of the membrane. Nevertheless, the polysilicon area is small in our design in comparison with the whole membrane area and will not affect significantly the stresses balance in the membrane.

1.4 Thermal design

1.4.1 Basis of heat transfer theory

Three modes of heat transfer are to be considered: conduction, convection and radiation [11]. Each of them will be briefly summarized below.

Conduction

Conduction refers to heat transfer by diffusion through a solid material or non-moving fluid. It is analogous to electrical conduction [11]. When there is a temperature gradient present in a body, heat will flow from the hotter to the colder region, and the heat flux F (in W) will be proportional to the temperature gradient across the body, according to Fourier Law [13]:

$$F = \frac{dQ}{dt} = -kA\frac{dT}{dx} \quad (1.1)$$

where Q = quantity of heat transfered in Joules (J); k = thermal conductivity, in W/m K; A = cross-sectional area, in m^2; T = temperature, in K and x = distance in the direction of heat flux normal to A, in m. The minus sign means that the heat flux is moving in an opposite direction of the gradient. Heat flux, is analogous to the "current" of heat flowing across a unit area per unit time and thermal conductivity can be related to the electrical conductivity. The more general form of this equation is an expression of the vector of heat flux density in W/m^2, as

$$\vec{q} = -k\nabla T \quad (1.2)$$

where ∇ is the three-dimensional del operator. It can therefore be expected that conduction losses will depend on the size (x), thickness (A) and material of the membrane (k), as will clearly be observed in the following.

More interesting is the diffusion equation [11] which represents the temporal diffusion of thermal energy as follows:

$$\frac{dT}{dt} = \frac{k}{\rho c}\nabla^2 T = \alpha \nabla^2 T \qquad (1.3)$$

where ρ is the mass density, in kg/m^3, and c is the heat capacity, in J/kg K. The heat capacity of a body can be defined as its ability to "hold" thermal energy, in a manner analogous to electrical capacitance. Finally, α is often refered to the thermal diffusivity of a material (in m^2/s). This parameter reflects the thermal conductivity of a material relative to its ability to store thermal energy (analogous to an electrical RC time constant).

The thermal properties of thin-film materials often differ significantly from those published for bulk samples. Furthermore, variations between runs of a single process can be non negligible [11] and could require the use of dedicated test structures for characterization on each run as proposed in [112][113]. In our case, only thermal conductivity will be useful to perform our simulations. Its value for each constitutive layer of our membrane was found in literature and a theoretical global value for our membrane was calculated and fits our measurements.

Convection

Convection refers to heat transfer by the movement of fluid or gas. It is very difficult to model numerically and is generally estimated by empirical methods [11]. The heat transfer by convection between a fluid at temperature T_f and a surface at temperature T_s can be expressed by the Newton law :

$$F = Ah(T_p - T_f) \qquad (1.4)$$

where F is the heat flux in W, A it the area in m^2 and h is the emperitical convection coefficient in W/m^2 K. This last coefficient depends on a lot of factors and can not be fixed for a given geometry.

Two types of convection can be found: free and forced. Free convection occurs when a fluid moves and transfers heat due to thermal gradients. In the other hand, we are in

1.4. THERMAL DESIGN

forced convection when the fluid transferring heat is externally pumped. In our case, only free convection between the top or the bottom surface of the membrane and air will occur. The convection coefficient used will be extracted from literature as it will be explained further.

Radiation

Radiation refers to heat transfer via the emission of electromagnetic waves [11]. Every body may exchange heat by emitting or absorbing thermal radiation [13]. In our case, the radiation transfer is found insignificant as confirmed in [114][76][100][106].

1.4.2 Thermal simulations

ANSYS simulations were performed to optimize the microheater size and shape as well as the membrane area and thickness to ensure a uniform temperature distribution over the whole active area and to achieve low power consumption.

The conventional meander, or double meander design found in most devices, covers the whole heated area and creates a central hot spot and a high temperature gradient from the center to the border of the active zone [99][88][100][115][19][110] as shown in Fig. 1.2(a).

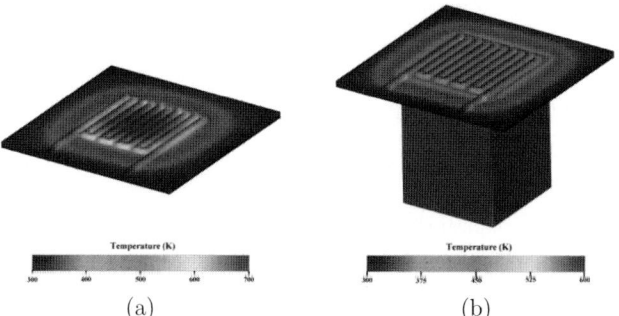

(a) (b)

Figure 1.2: Classical meander introducing hot spot in the middle of the membrane, (a) without and (b) with a silicon heat spreader under the active are (from [88]).

Hot spots can produce burn-through and reduce the sensor lifetime [116]. Hot spots also decrease the sensitivity as well as the discrimination towards the target gas. To compensate this lack of uniformity, a lot of publications often introduce either a silicon island underneath the active area [99][88][100][115][105][97][117][110] (Fig. 1.2(b)), or a thermal conductive layer such as aluminum over the structure [91][89] or polysilicon underneath the dielectric membrane [118]. This improvement increases the robustness of the membrane but also leads to higher thermal inertia, higher consumption as well as higher process complexity. Furthermore, aluminum must be avoided if high temperatures are required.

More recently, [119] reported an original design consisting of two loop-shape microheaters, one smaller being confined into a larger one parallel to the perimeter of the membrane. It is shown that the independent operation of both heaters is favorable for a better temperature distribution compared to the operation of only one heater.

Our idea was to replace the classical meander by a loop surrounding the active area we would heat as partially suggested in [100] but considered quite arduous to simulate [100]. Furthermore, our dielectric membrane would be only covered with a high thermal conductivity gas sensitive layer (metallic oxide layer like SnO_2 or WO_3) to increase the conductivity as well as the uniformity inside the heater perimeter without increasing the power consumption as confirmed by our simulations.

The literature [76][115][19] indicated that for a 1 μm thick specific dielectric membrane, a difference of 400°C appears over a distance of 200 μm. Assuming that our case can be simplified to a 2D problem of thermal conduction, we can consider that the thermal profile is linear in the membrane. Therefore, we could expect that the thermal profile would be constant at 0°C between two infinite resistor lines heated at 400°C and separated by 200 μm. Therefore, a typical size of 200x200 μm^2 was chosen as active area and a 200 μm wide loop resistor enclosing this area. In the same way, a minimum separation of 200 μm between the sensitive area and the edge of the membrane was required for achieving full thermal insulation on the device and maintaining the silicon at ambient temperature. Therefore, a minimum membrane dimension of approximately 600x600 μm^2 was needed. A trade-off appears here between high thermal insulation and membrane robustness. For operation at temperatures up to 450°C, we have chosen a typical membrane size of 640x640 μm^2 for an active heated area of around 240x240 μm^2. But two other membrane sizes, 440x440 μm^2 and 840x840 μm^2 were also tested. The impact of the membrane size on the power consumption will be discussed in details in the experimental subsection.

These first considerations helped us to build a more accurate design by ANSYS numerical finite elements simulations. As the membrane and the loop are symmetrical,

1.4. THERMAL DESIGN

simulations were performed only on a quarter of the membrane to simplify the calculation time assuming that thermal conditions are adiabatic on each symmetrical edge. It is consistent as the heat flux is perpendicular to each edge.

The other boundary conditions were fixed as follows: the temperature at the periphery of the membrane is constant and fixed equal to the ambient temperature (30°C) [76]; on the upper and lower surfaces of the membrane, the heat is dissipated through convective exchange with the gaseous phase. The exchange coefficients were taken as 250 W/m² K and 125 W/m² K for the front and back sides of the membrane, respectively according to experimental measurements in vacuum reported in literature [114][76]. Finally, vertical temperature gradient and vertical conduction are neglected and the design can therefore be advantageously simplified in a 2D model [73]. Radiation losses were also considered as negligible [114][76][100].

The heat generation was set by a power generator in the resistor which was calculated as follows,

$$Injected\ Power\ density = \frac{P_{microheater}}{V_{microheater}} \quad (1.5)$$

where $P_{microheater}$ is the power consumption of the microheater and $V_{microheater}$ is the volume of the heater. The volume of the heater is fixed by its thickness (340 nm), its length (600 μm for the 3 sides and 400 μm for both accesses) and its width (20 μm). Therefore, to reach a typical power consumption of about 25 mW, the power density to inject in the loop was fixed at around $0.53*10^{11}$ W/m³.

Table 1.1: Thermal parameters of membrane materials and of main others microelectronics materials as comparison.

Material	Density [kg/m³]	Conductivity k [W/m K]	Thickness [nm]	Ref.
Si	2330	50...150[*]	x	[73][114][120]
PolySi	2330	18[**]	x	[73][114][121]
SiO₂	2200	1.38	400	[73][114][76][120]
Si₃N₄	3200	22.7	300	[73][114]
SnO₂	2500	15	300	[27]
Air	-	0.024	-	[120]

[*] depends if silicon is doped or not
[**] if doped at 10^{20} at./cm³. For more informations about conductivity of doped PolySi layers, see [121]

The theoretical conductivity k_m value of the three stacked dielectric membrane of total thickness t_m can be calculated by superposition as follows [122],

$$k_m = \frac{k_{ox}t_{ox1} + k_{nit}t_{nit} + k_{ox}t_{ox2}}{t_m} \qquad (1.6)$$

where k_{ox} and k_{nit} are the conductivities of oxide and nitride, respectively and t_{ox1}, t_{nit} and t_{ox2} are the thicknesses of each constitutive layer of the membrane. All values are summarized in Table 1.1. The theoretical conductivity k_m was calculated equal to 7.6 W/m K. But we can expect an actual lower value for the stacking than for the bulk material due to surface and interface defects. The literature [76] suggests values between 3.5 and 6 W/m K. We will see further that our measurements show a good agreement for a thermal conductivity of 5 W/m K. This value was used in our simulations and was confirmed by the literature [76]. However, it must be noted that polysilicon locally increases the conductivity and that gas sensitive layer (such as SnO_2) increases the conductivity of the whole membrane.

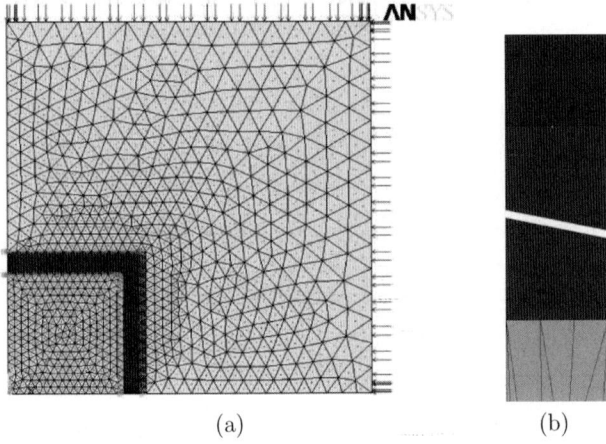

(a) (b)

Figure 1.3: (a) meshing of the upper right quarter (320x320 μm^2) of the membrane. The red arrows show the membrane borders. (b) meshing of the membrane section showing, from the bottom to the top, the three-stacked dielectric membrane, the polysilicon layer and the metallic oxide layer.

The meshing was wisely fixed to be continuous on the whole membrane quarter. It means that each node of each surface (resistor, upper membrane area, lower membrane area) must be joined on the border of the volumes. The problem is difficult in this case since ANSYS meshes each volume separately. Fig. 1.3(a) shows the meshing of a quarter

1.4. THERMAL DESIGN

of the membrane. We can observe the meshing refinement on the heater border to provide continuous junctions. Fig. 1.3(b) shows the vertical meshing in a shrinked section of the membrane with its different layers.

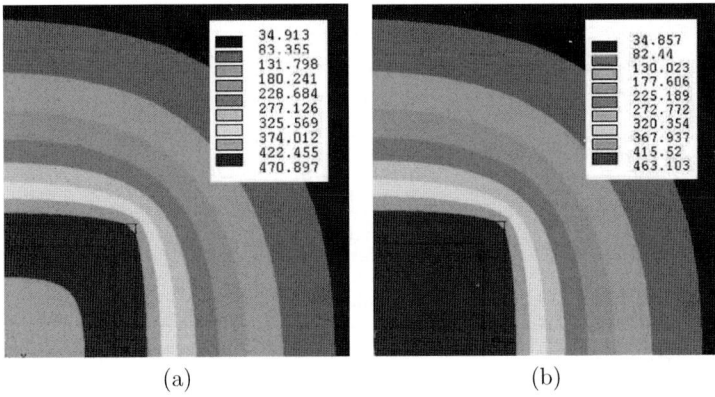

Figure 1.4: Thermal uniformity (Temperature in °C of a quarter (320x320 μm^2) of membrane without (a) and with 1 μm thick SnO_2 layer on top of the membrane (b). The straight lines locate the loop shaped polysilicon resistor (injected power = 25 mW).

Final results can be shown in Fig. 1.4(a) and (b), respectively without and with a 1 μm thick SnO_2 film as sensitive layer on top of the membrane. In both cases, the power consumption is fixed around 25 mW. A temperature variation of maximum 50°C can be observed on the active area when not covered with the metallic oxide layer. These results confirm that metallic oxide layer increases the thermal uniformity of the membrane thanks to its high thermal conductivity.

The effect of the membrane thickness on the power consumption and thermal uniformity has also been demonstrated by simulations (Fig. 1.5). If the membrane thickness is increased by a factor 2, the membrane conductivity remains constant as demonstrated by Eq. 1.6 if t_{ox1}, t_{nit}, t_{ox2} and t_m are doubled in same time. However, the thermal uniformity into the membrane is improved because of the better thermal conduction achieved thanks to the cross section area increase (see Eq. 1.1). Unfortunately, the power to achieve 400°C is increased by around 20 %. This slight power consumption increase is also observed by experiments in the literature [73][76]. The heat lost by conduction through the membrane thickness is negligible in comparison with the heat lost by conduction through the surface [73]. In silicon membrane sensors, it is the contrary, the power consumption is strongly related to the membrane thickness [76].

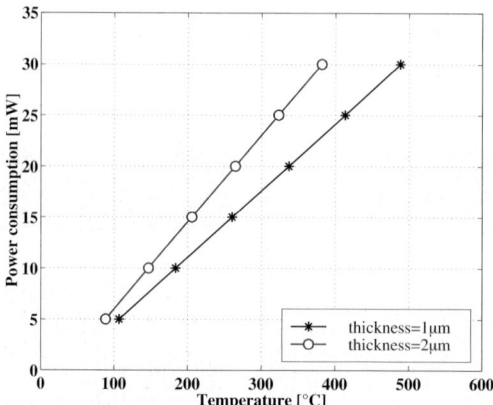

Figure 1.5: ANSYS simulations showing the influence of the membrane thickness on the power consumption versus membrane temperature with a thermal conductivity of 5 W/mK.

On the contrary, a 1 μm thick membrane leads to a decrease in power and in thermal uniformity but also in mechanical robustness. This confirms that a membrane of 1 μm of thickness offers a good trade-off between uniformity, power consumption and mechanical resistance thanks to the dielectric sandwich to compensate the residual stresses. A thickness of 1 μm is also imposed by our technology since the first layer of our membrane (the SOI buried oxide) is fixed to 400 nm.

1.5 Device fabrication

1.5.1 On bulk silicon substrate

To demonstrate the microheater prototype, we firstly emulated our SOI process on bulk silicon substrate. The starting material was a $\langle 100 \rangle$ silicon p-type 200 μm double sided polished wafer (Fig. 1.6). The two first layers of the membrane were 400 nm grown SiO_2 to simulate the buried oxide of a SOI wafer and 300 nm LPCVD Si_3N_4 (SiH_2Cl_2/NH_3 ratio equal to 3). A 340 nm layer of polysilicon was then deposited by LPCVD at 625°C in a SiH_4(Silane) atmosphere (deposition rate of around 90 Å /min). Its doping was performed with phosphorous impurities in solid phase at 900°C during 1 hour to saturation for obtaining a sheet resistance around 27 $\Omega/\square$. Polysilicon was finally patterned and a 300 nm thick PECVD oxide layer was deposited and densified to constitute the third layer

1.5. DEVICE FABRICATION

of the membrane. Via holes were then opened and a 900 nm thick aluminum (+2 % Si) layer was sputtered and patterned to create bonding pads and to contact polysilicon.

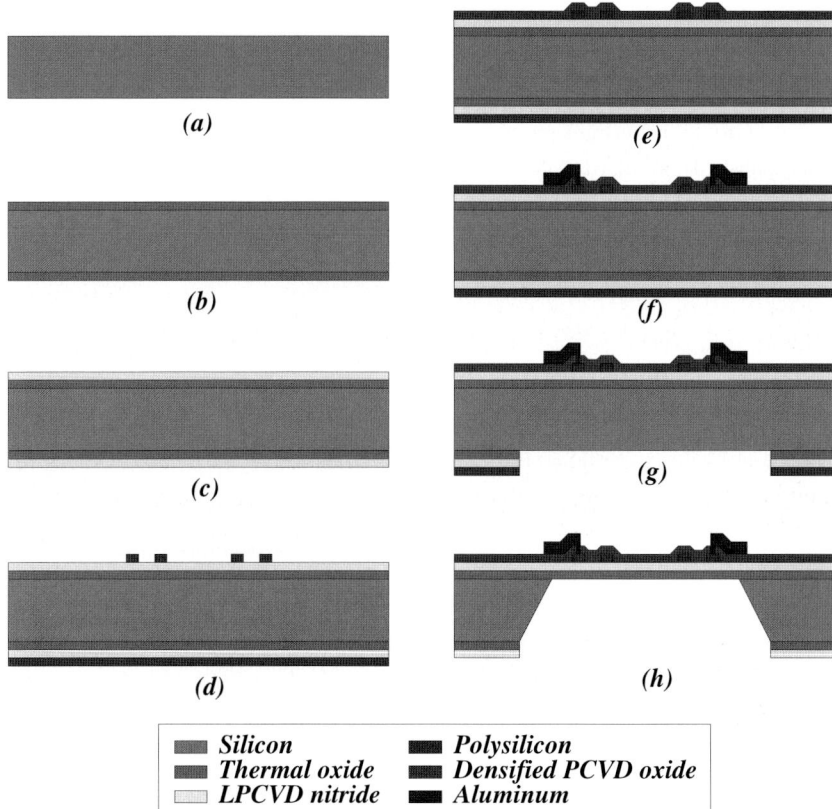

Figure 1.6: Schematic process flow of a microheater in bulk technology: (a) 200 μm Si wafer, (b) thermal oxidation, (c) LPCVD nitride deposition, (d) polysilicon deposition and patterning, (e) PECVD oxide deposition and densification, (f) aluminum deposition and patterning, (g) backside window patterning and (h) TMAH micromachining.

The backside window was then defined using the double side mask alignment and the three layers mask (polysilicon, nitride and thermal oxide) previously deposited on the backside of the wafer, for silicon protection in TMAH, were patterned by plasma RIE. Finally, the membrane was created by a single post-processing step, consisting of a backside bulk micromachining. Anisotropic etch was performed with TMAH and stopped by thermal oxide in conditions explained in the first chapter (part one) to achieve high selectivity versus aluminum.

1.5.2 On SOI substrate

In case of SOI-CMOS technology, the starting material was a SOI UNIBOND 500 μm (one-side polished) standard wafer with a buried oxide of 400 nm and a 100 nm thick silicon film (Fig. 1.7(a)).

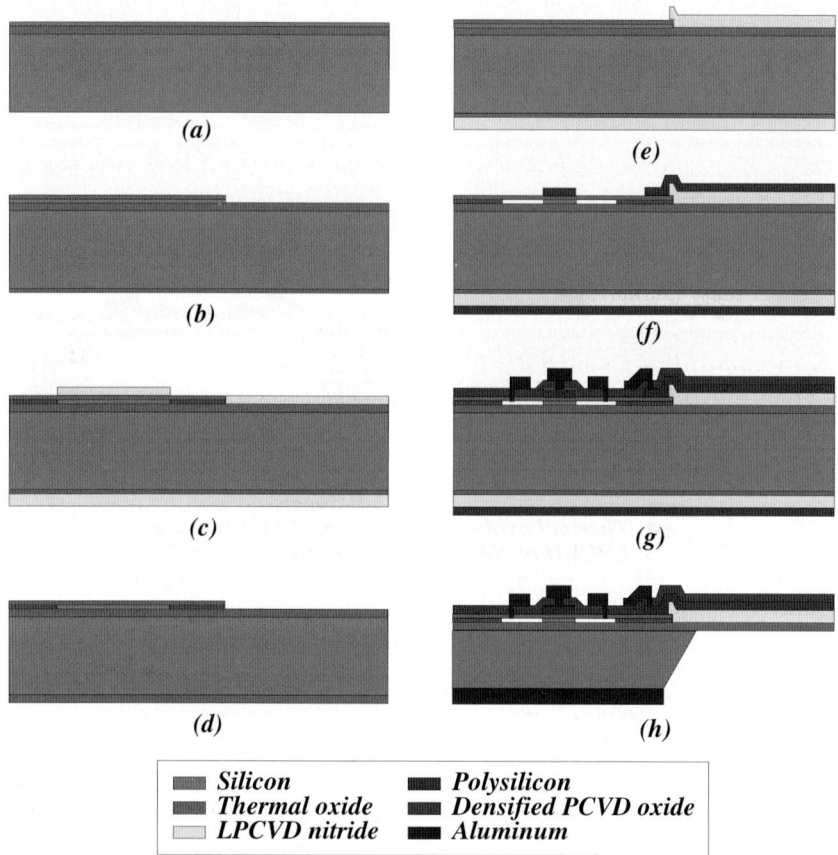

Figure 1.7: Schematic process flow of integrated SOI-CMOS microheater. A SOI-CMOS circuit is realized (in left part) jointly with the microheater on membrane (in right part).

1.5. DEVICE FABRICATION

After an initial oxidation, the silicon film was etched by plasma throughout its whole thickness at the future membrane location (Fig. 1.7(b)). A standard LPCVD nitride layer of 200 nm thick was then deposited and patterned to protect the electronics active zones from the LOCOS field oxidation (Fig. 1.7(c)). After LOCOS, nitride was fully striped (Fig. 1.7(d)) and replaced by a thicker LPCVD nitride layer of 300 nm (Fig. 1.7(e)) to implement the second layer of our three stacked dielectric layers membrane with the 400 nm thick SOI buried oxide. Standard CMOS channel implantations were then performed followed by the gate polysilicon deposition and patterning and by the standard source and drain self-aligned implantations (Fig. 1.7(f)). Then, a 300 nm PECVD oxide was deposited and densified to constitute the third layer of the membrane and the interconnection insulating layer of our CMOS circuits. Via holes were opened in the oxide layer and 900 nm of aluminum (+2% Si) were sputtered and patterned to contact polysilicon and silicon doped regions (Fig. 1.7(g)).

Finally, the backside wafer was thinned and mechanically polished to reach a thickness of 200 μm, more appropriate to a silicon wet etching. A 500 nm aluminum layer was then evaporated on backside before its patterning using double side alignment. This step was followed by TMAH etching such as explained in previous part.

We can observe that outside of the membrane, the upper monocristalline silicon film is advantageously used to integrate reliable electronics close to the microheater at the same time. On the same way, the SOI buried oxide layer is used as the first layer of our membrane as well as etch stop for TMAH etching. Furthermore, aluminum on backside is advantageously used as back gate contact for integrated circuits and as mask to protect silicon during TMAH etching. In bulk technology, this co-integration could be performed using the LOCOS thermal oxide as the first constitutive layer of the membrane. Nevertheless, its thickness would have to be controlled with high accuracy in this case.

Fig. 1.8, 1.9 and 1.10 respectively depict the microheater on membrane in CMOS-SOI technology, a close view of the transition between the circuit part (LOCOS oxide) and the sensor part (the membrane location) and a cross section of the transition. Note that interdigitated gold electrodes are integrated on the membrane (Fig. 1.8) for gas sensing. We must especially note the little overlap of the LPCVD nitride (4 μm) on the LOCOS oxide at the boundary of the two parts to prevent any lack of precision during the photolithography alignments. We will come back to this process in the Chapter 4 of this part when we will speak about the integration of transistors on membrane.

In comparison with the standard SOI-CMOS process, only 2 more masks are needed

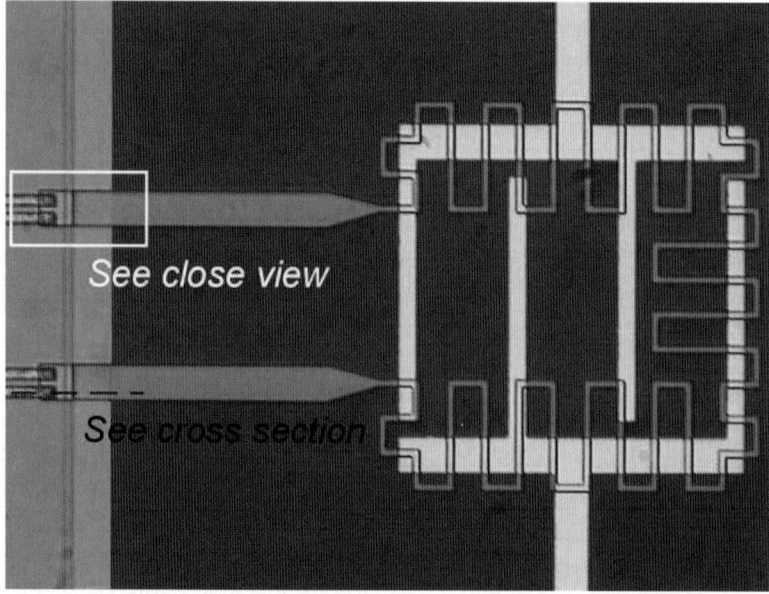

Figure 1.8: Detail of a microheater on membrane in CMOS-SOI technology. We can observe the interdigitated gold electrodes on the microheater for gas sensing (see close-view in Fig. 1.9 and cross-section in Fig. 1.10).

1.5. DEVICE FABRICATION

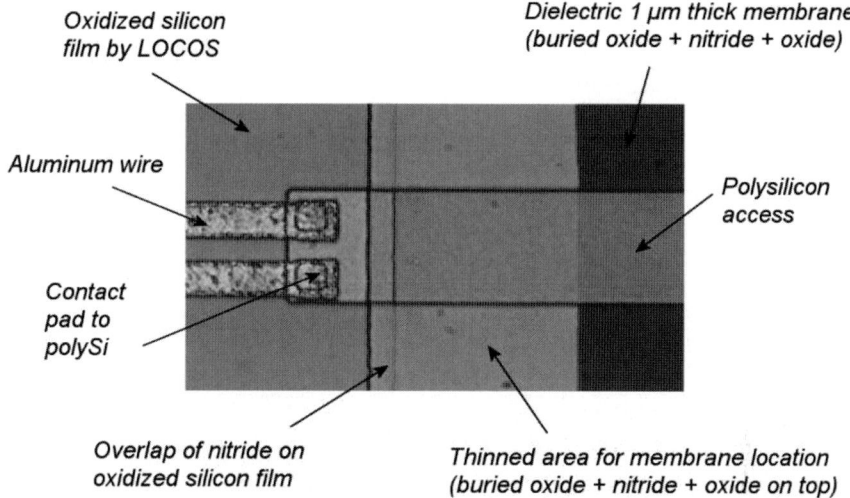

Figure 1.9: Close-view of the transition between the circuit part (LOCOS oxide) and the sensor part (membrane location) in SOI-CMOS technology.

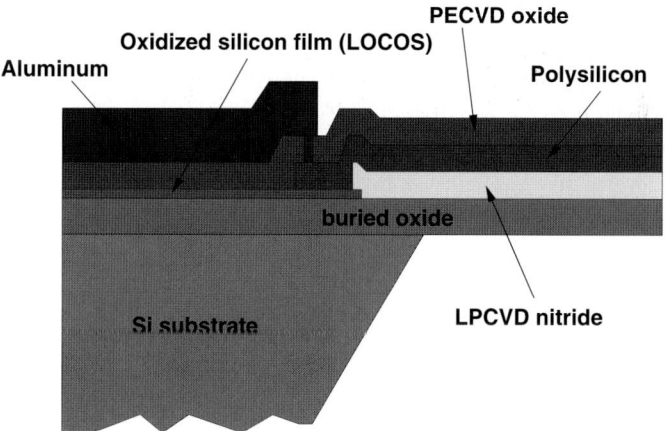

Figure 1.10: Cross-section of the transition between the circuit part (LOCOS oxide) and the sensor part (membrane location) in SOI-CMOS technology.

(if we do not take into account the post-processing backside etching): the first one to pattern silicon on the membrane location (Fig. 1.7(b)) and the second one to etch the nitride layer (Fig. 1.7(e)). This process therefore also requires 2 more photolithography steps, 1 more deposition (LPCVD nitride layer) and 2 more etching steps (silicon film and nitride layer). Each thickness is the same except the densified PECVD oxide which is a little bit increased in comparison with the standard interconnection layer (300 nm instead of 250 nm). Its thickening is required to adjust the residual stresses in each layer as well as to reach the target thickness for our membrane. This thicker layer plays also the role of passivation layer for the polysilicon film.

1.5.3 Dicing and packaging

Wafers at the end of the backside micromachining are very fragile and need to be handled with care to avoid any breaking. Back etch is one of the critical process steps concerning mechanical yield. After TMAH etching, wafers were immersed in boiling DI water for a better rinsing and dryed in methanol to avoid damaging membranes in standard rinser dryer. We observed a high yield at this step since statistically, on 9 wafers featuring more than 150 membranes/wafer, only between zero and two 800x800 μm^2 or 600x600 μm^2 membranes/wafer were broken.

Some post processing steps were carried out on two wafers to check the behavior of our membrane when we imposed them more mechanical stress. By this way, we performed tests which consisted in removing aluminum for replacing it by a new layer (as it was needed when aluminum was damaged in TMAH). Aluminum stripping in H_2SO_4, followed by aluminum sputtering, photolithography (photoresist spinning, backing on hot chuck, exposition, development and rinsing in rinser dryer) and finally aluminum etching in plasma were achieved. These aggressive tests did not decrease too much the yield since only 1 or 2 more membranes were broken. We also observed that the more critical step was the photoresist spinning for which the wafer was held by vacuum on the chuck introducing a lot of stress in our membranes. A more appropriate technique would consist to use a chuck maintaining the wafer by 4 lugs without vacuum. On the same way, gas sensors post processing was also currently used to deposit sputtered sensitive layer on membrane, and revealed similar yield of only 1 more membrane broken at the end (observed on 3 wafers).

Wafer dicing was another source of yield loss. Standard microelectronics sawing was preferred to the diamond scriber to decrease the induced stresses and to avoid the depo-

1.6. MICROHEATER CHARACTERIZATION AND RESULTS

Figure 1.11: Packaged microheater.

sition of filth on membranes. Micromachined wafers were previously taped from backside on a sticky blue foil and sticked on the chuck of the saw to prevent unsticking and damages when vacuum was applied. During sawing, the pressure of the DI water used to cool the saw was adjusted to avoid damages. At the end of the dicing, chips were manually pulled away from the blue foil. In the future, this step will be performed easier using a dedicated thermo sticky blue foil which can be removed by a simple heating. Chips were finally bonded on standard DIL ceramic package using epoxy glue. Some membranes were broken after dicing. Nevertheless, we observed a higher yield for membranes patterned on the middle of the chip.

1.6 Microheater characterization and results

1.6.1 Introduction

Four configurations have been considered to study in details the effects of the loop and access lines geometry on the power consumption (Fig. 1.12). The four metallic contacts we can see in Fig 1.12 allow a four-point measurement of resistance (and then temperature) simultaneously with Joule heating. Meanders were introduced in the three first configurations to increase the loop resistance and therefore increase the thermal conductivity. The third has a larger meander period than the two first ones and the fourth configuration has the classical shape as a control. The difference between the first and the second configurations is the width of the polysilicon access lines to measure their impact on the power consumption.

The dimensional parameters of the loop and accesses are summarized for each con-

CHAPTER 1. LOW POWER MICROHOTPLATE AS BASIC CELL

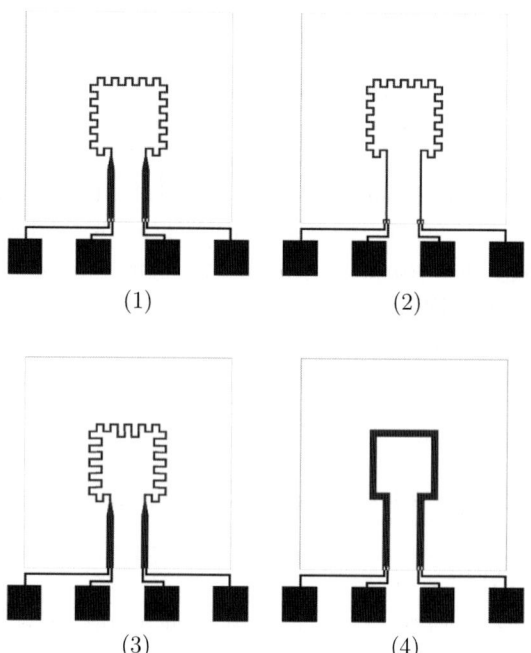

Figure 1.12: The four configurations of loop shape polysilicon resistor.

1.6. MICROHEATER CHARACTERIZATION AND RESULTS

figurations in Table 1.2. The loop length (L_{loop}) and the loop section (S_{loop}) are showed as well as the access length (L_{acc}) and their section (S_{acc}). The resistance values of the loop and the accesses were calculated from the polysilicon sheet resistance (fixed by the CMOS process between 27 and 30 $\Omega/\square$) and polysilicon thickness (340 nm). The polysilicon resistivity was extracted by multiplying its sheet resistance in $\Omega/\square$ by the polysilicon thickness in $10^6 nm$. Resistivity values between 9.18 and 10.2 $*10^{-6} \Omega m$ were obtained. Finally, R_{tot} calculates the sum of the loop and access resistances.

Table 1.2: Dimensional parameters and calculated resistances of the loop-shaped microheater.

Config	L_{loop} [μm]	S_{loop} [μm^2]	L_{acc} [μm]	S_{acc} [μm^2]	R_{loop} [Ω]	R_{acc} [Ω]	R_{tot} [Ω]
(1)	1560	5x0.34	400	20x0.34	8430-9360	540-600	8970-9960
(2)	1560	5x0.34	400	5x0.34	8430-9360	2160-2400	10590-11760
(3)	3126	5x0.34	400	20x0.34	16880-18760	540-600	17420-19360
(4)	781	20x0.34	400	20x0.34	1055-1172	540-600	1595-1772

The following lines will describe the measurements we carried out on microheaters in configuration 1. Only in the subsection "discussion on the microheater geometry", we will compare the four configurations to enlarge the study.

1.6.2 Calibrations

Our microhotplate uses a polysilicon resistor as microheater as well as temperature sensor, based on the ability to calibrate its resistance as a function of temperature, and then uses an in situ measurement of the resistance to determine its temperature during operation. Before the heating measurements of our microheater, it was therefore needed to measure the polysilicon resistance variation versus the temperature. This step is in fact the calibration of the heater to foresee in our next measurements the temperature reached at a given current or voltage applied.

Resistance measurements were performed by heating the membrane on a hot chuck when applying a very low power <1mW (current of 10 μA) across the resistor to avoid Joule heating. Measurements were carried out at 5 temperatures between 22°C and 250°C and were extrapolated until 500°C as shown in Fig. 1.13. It is reported in [123] that the change in resistance versus temperature can be extrapolated up to 700°C with good accuracy. This assumption will be confirmed further by our measurements at higher temperatures. The temperature of the hot chuck was controlled to within ±1°C during these calibrations.

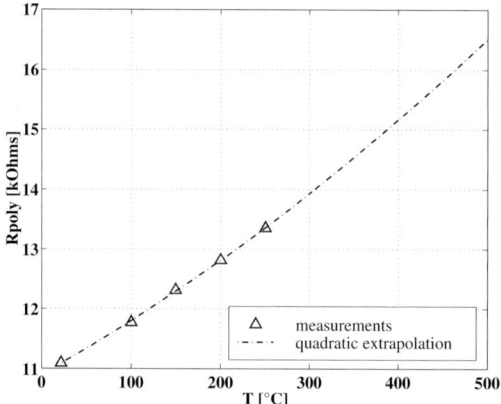

Figure 1.13: Calibration of the microheater: Polysilicon resistance variation versus applied external temperature.

Fig. 1.13 shows that the resistance variation with temperature is slightly quadratic or quasi linear and that the thermal conductivity decreases if the temperature increases (positive slope), as in any conductor. It means that when the temperature increases, the power dissipation increases more in a hotter area than in another one and thus creates hot spots. A negative slope (in insulators or semiconductors for instance) would be better to have a more uniform temperature distribution but requires a lower doping level of polysilicon, and thus a higher resistance and a higher power consumption. In our case, a polysilicon sheet resistance of around 27 Ω/□ imposed by our IC process is a good trade-off between good thermal uniformity and low power consumption for microheaters.

Calibrations were performed on more than 60 microheaters of the same configuration (config 1), across the same wafer, from wafer to wafer in a same run and between two batches to study their dispersion. To quantify this dispersion, the slope of the curve as well as the temperature coefficient of resistance (TCR) were evaluated for a given temperature, chosen at 200°C, at the middle of the temperature range. TCR (in °C^{-1} or in %/°C) at 200°C was calculated as follows [124][125]:

$$TCR = \frac{1}{R_{200}} \left| \frac{dR(T)}{dT} \right|_{T=200°C} = \frac{1}{R_{200}} * Slope \qquad (1.7)$$

1.6. MICROHEATER CHARACTERIZATION AND RESULTS

where R_{200} is the resistance at 200°C and $\frac{dR(T)}{dT}$ is the slope of the resistance-temperature curve at 200°C (termed in $\Omega/°C$). The TCR is theoretically constant through a uniformly doped polysilicon film. On the contrary, the slope of the resistance-temperature curve is dependent on the resistance value at 200°C and therefore on the resistance value at ambient temperature.

As the resistance R_{poly} (in Ω) increases quadratically with the temperature T (in °C), we can express its variation with the following quadratic polynomial,

$$R_{poly}(T) = aT^2 + bT + c \quad (1.8)$$

The slope $\frac{dR(T)}{dT}$ and the TCR were therefore calculated from the derivative of the polynomial (Eq. 1.8) at temperature $T = 200°C$ as follows,

$$TCR = \frac{Slope}{R_{200}} = \left|\frac{2aT + b}{R_{200}}\right|_{T=200°C} \quad (1.9)$$

TCR measurement results on our 60 devices are depicted in Fig. 1.14(a) versus the sheet resistance previously measured on each wafer. Sheet resistance measurements were carried out during the process just after the polysilicon doping using a 4-points probe. Its value is therefore proportional to the phosphorus doping level. Fig. 1.14(a) shows that the TCR slightly increases when polysilicon resistivity increases, i.e. when doping level decreases as confirmed in literature [27]. This graphic also reveals that the same microheater design can show a dispersion in TCR values up to 10 % from one site to another site of the same wafer due to non uniformities in the doping by solid source. A TCR value around 0.08 %/°C was extracted which agrees with values found in the literature [126][123].

Since TCR variations were observed even across a same wafer, the polynomial expression (1.8) is slightly different from one microheater to an other and needs therefore to be recalculated for each microheater. Of course, it could be extracted by a separate calibration of each sensor. But more efficiently, it could be foreseen for any microheater by interpolation of our 60 microheaters measurements results. For this purpose, we extracted the variation of each polynomial coefficient a, b and c versus the room temperature resistance and we obtained the graph of Fig. 1.14(b). Since the results showed a monotonic variation, a linear interpolation was performed to be able to calculate the polynomial coefficients from any room temperature resistance. The following linear polynomials were

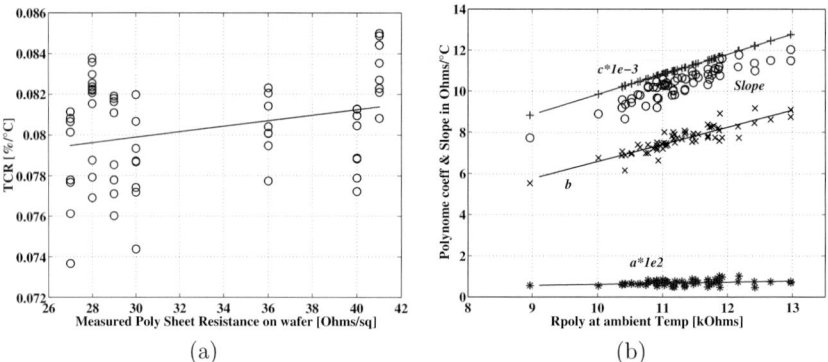

Figure 1.14: (a) TCR variation versus polysilicon sheet resistance; (b) variation of the polynomial coefficients and the slope of the resistance-temperature curve versus the polysilicon resistance value at ambient temperature.

found for each coefficient,

$$a = 0.00051 * R_{22} + 0.000994 \tag{1.10}$$

$$b = 0.8303 * R_{22} - 1.7157 \tag{1.11}$$

$$c = 980.6316 * R_{22} + 38.9362 \tag{1.12}$$

where R_{22} is the room temperature resistance (usually 22°C) and is expressed in $k\Omega$. The very low value of a demonstrates the slight quadratic behavior of the curve shown in Fig. 1.13. Fig. 1.14(b) shows also the evolution of the resistance-temperature curve slope calculated from Eq. 1.9. As explained above, it increases linearly, following the resistance room temperature resistance.

Since our numerous measurements are sufficiently representative of any microheater having the same design, we can now foresee the current-temperature behavior of a microheater knowing only its room temperature resistance R_{22}. Indeed, from R_{22}, the polynomial coefficients a, b and c can be calculated to extract the quadratic polynomial $R_{poly}(T)$. From this last polynomial, the temperature value can be extracted from any measured polysilicon resistor value.

The interpolation in the calculation of the coefficients a, b and c leads to an error of maximum 5-6 % on the final extracted temperature. This value was calculated from the two measured points featuring the highest deviation with respect to the linear interpolation (see the coefficient "b" in Fig. 1.14(b)). The error means an acceptable inaccuracy

1.6. MICROHEATER CHARACTERIZATION AND RESULTS

on the calculated temperature of only 20°C around 400°C.

By this way, our calculated polynomial avoids to perform tedious calibrations on each microheater separately for a given design and process.

1.6.3 Measurements

To measure the current-voltage relation of the microheater, an increasing current was applied on two bonding pads while the voltage was measured on the others to obtain a typical curve as shown in Fig. 1.15(a).

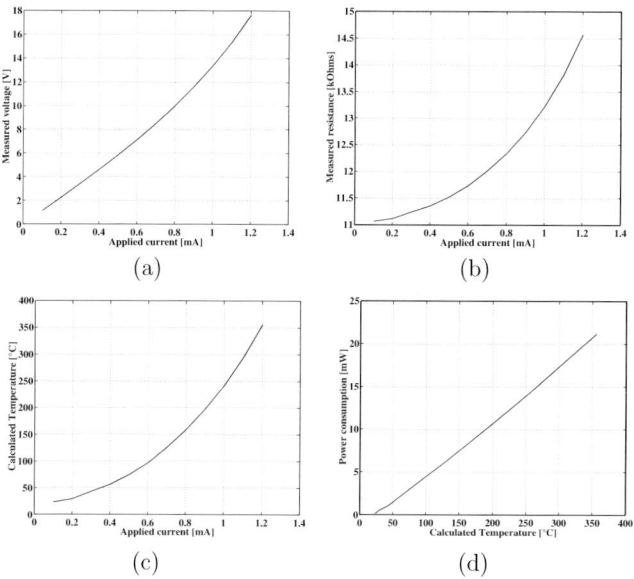

Figure 1.15: Successive results of measurements and calculations from the current-tension curve towards the variation of the power consumption versus the target temperature (config 1 and membrane size of 840x840 μm^2).

The resistance was then calculated for each current (Fig. 1.15(b)) and using our calculated polynomial (Eq. 1.8), the temperature was calculated for each current (Fig. 1.15(c)). Finally, the power consumption was calculated according to the microheater temperature (Fig. 1.15(d)).

The effect of membrane dimensions has been thoroughly investigated. power consumption slowly decreases with larger membranes since the thermal insulation is more efficient in this last case (Fig. 1.16). Therefore, larger membranes can be useful to work at higher temperature eliminating the larger thermal flux to the substrate. Nevertheless, larger membranes are more fragile. There is therefore a trade-off between robustness and power consumption for the intermediate membrane (640x640 μm^2).

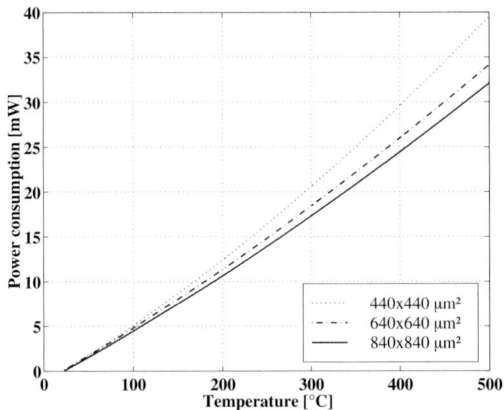

Figure 1.16: DC Power consumption vs. temperature elevation measurements for 3 different sizes of square membranes (840 μm, 640 μm and 440 μm on one side) supporting the same resistor configuration (config 1).

Performance of the heater can be judged in terms of power consumption versus temperature elevation. As shown in Fig. 1.16, the active area of our membrane can be heated up to 400°C with less than 25 mW. This result outperforms most recent CMOS compatible realizations as summarized in Table 1.3. Our excellent results are mainly related to the comparatively small total thickness of the membrane. Indeed, our simulations showed that the power consumption increased with the membrane thickness (Fig. 1.5).

It can be noted that some of the realizations presented in Table 1.3 were reported able to be heated up to 700°C. In a next section, we will discuss about this interesting characteristic.

Finally, the comparison between simulations and experimental results (on a 640x640 μm^2 membrane) revealed a good agreement between measurements and simulated values with a thermal coefficient of 5 W/m K as shown in Fig. 1.17. This conclusion confirms our previous assumption.

1.6. MICROHEATER CHARACTERIZATION AND RESULTS

Table 1.3: Comparison of our design with recent published microheaters results.

Group	Year & Ref.	Power consumption	Heated area [μm^2]	Mem. thick. [μm]	Heater mat.	Membrane material
CNM	1997 [19]	50 mW @ 350°C	500x500	2	polySi	Si_3N_4/SiO_2
LAAS	1998 [73]	50 mW @ 230°C	100x2000	0.7	polySi	$SiO_2/SiN_{1.2}$
Motorola	2000 [114][76]	65 mW @ 450°C	430x600	1.5	polySi	SiO_xN_y
Samlab	2000 [99][88]	55 mW @ 300°C[1]	500x500	1	Pt	Si_3N_4/SiO_2
INFM	2001 [127][128]	30 mW @ 400°C[1]	?	1.5	polySi[2]	Si_3N_4/SiO_2
IMEL	2002 [129]	25 mW @ 700°C[1]	100x100	4	Pt	porous Si[2]
DICE	2002 Thesis	27 mW @ 400°C[1]	240x240	1	polySi	Si_3N_4/SiO_2

[1] can be heated up to 700°C
[2] not CMOS compatible due to use of not compatible layers

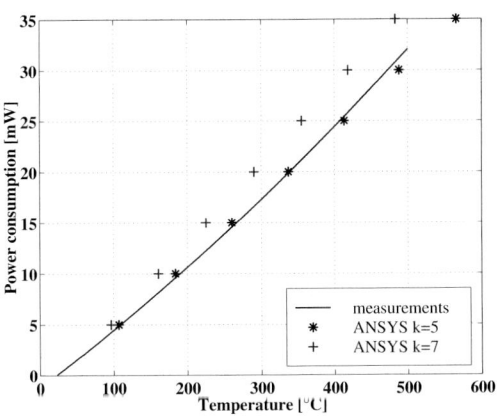

Figure 1.17: DC Power consumption versus temperature elevation. Comparison between simulated (for k=5 and k=7) and measured values for a 640x640 μm^2 and 1 μm thick membrane (config 1).

1.6.4 Discussion on the microheater geometry

Measurements were performed on the different microheater configurations for each membrane size to study the impact of the microheater geometry on the power consumption.

Fig. 1.18(a) and (b) show the measured slope and TCR values for each configuration and each membrane size. These results confirm that the TCR is of course independent on the configuration and the membrane size as well as to the presence or not of a membrane. As observed before, the slope increases with the resistance value at ambient temperature. This behavior is interesting since it conditions how fast (regarding the current increasing) a resistor will reach a target temperature and which power it will consume.

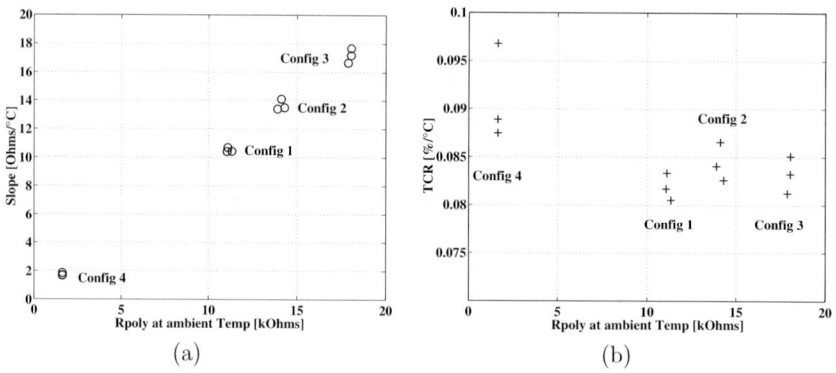

Figure 1.18: (a) Slope and resistance values for each configuration and each membrane size and (b) dispersion of the TCR value from one configuration to an other.

Taking these TCR values into account, the four configurations were compared in term of power consumption as depicted in Fig. 1.19 for a 840x840 μm^2 membrane size. As expected, no difference was observed between configurations 1 and 3 despite of the difference in their loop resistance values. The resistance increase between configurations 1 and 3 is compensated by its proportional slope increasing. Since config 3 has a higher slope, it reaches a given temperature faster than config 1 with a lower current and a higher resistance value. Therefore, both configurations consume the same power at this given temperature. On the contrary, the difference between configurations 1 or 3 and configuration 2 is due to their different access lines.

Access lines are responsible for the main part of the power consumption and thermal flux to the substrate. In configurations 1 and 3, the accesses have a very low resistance

1.6. MICROHEATER CHARACTERIZATION AND RESULTS

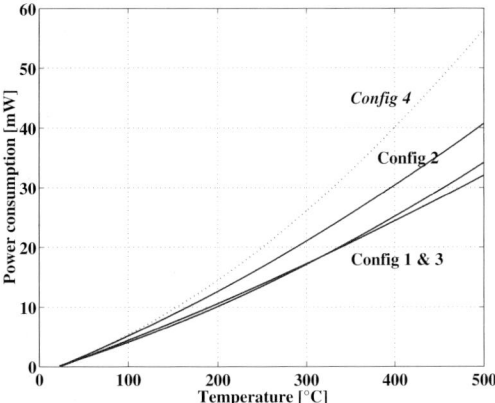

Figure 1.19: Comparison of the measured power consumption versus temperature elevation for each configuration (membrane size 840x840 μm^2).

in comparison with the heating loop, respectively 6 % and 3 %. On the other hand, configurations 2 and 4 feature a higher proportion, respectively 25 % and 50 %. About the geometry of the access lines, the area perpendicular to the flux is four times narrower in the second configuration than in the others (see Table 1.2). Therefore, the thermal flux to the substrate is lower than in the other configurations and when the loop heats, the resistance of the accesses does not increase significantly. Regarding the slope value in this configuration, it is in the average of configurations 1 and 3 and would not lead to a higher consumption. Nevertheless, the intrinsic higher resistance value of the accesses in this configuration needs to consume more power to heat the loop to a target temperature. It explains the slight power consumption increase we can observe in Fig. 1.18(b).

In fact, the best option would be to use the accesses geometry of configuration 2 to decrease their thermal dissipation while selectively increasing their doping level in order to decrease their resistance value. On the other hand, the doping level of the loop resistor would be unchanged to provide a high slope and therefore a lower power to reach a target temperature. Unfortunately, this option would require to replace our doping in solid source by implantation and would need one more photolithography mask.

In configuration 4, the power consumption increase is mainly due to the very low resistance of the loop. Its low resistance value leads to a lower slope than in other configurations. Therefore, more power is needed to reach a given temperature. In same time, the

138   CHAPTER 1. LOW POWER MICROHOTPLATE AS BASIC CELL

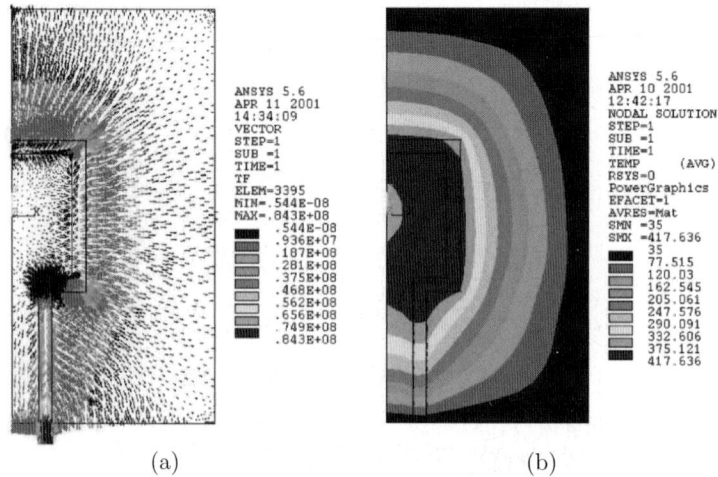

(a)   (b)

Figure 1.20: ANSYS simulation of a microheater in configuration 4 showing a high heat flow (a) and a high temperature dissipation (b) towards the accesses (power consumption = 25 mW).

temperature of the access lines increases and leads to an increase of their resistance value since they almost have the same resistance than the loop. In the other configurations, the accesses did not heat as much since they have a smaller resistivity. Consequently, more heat flux is driven towards the tip of the accesses and therefore towards the aluminum contacts as confirmed by simulations (Fig. 1.20). At high temperature, the aluminum contacts could reach more than 400°C and melt down. The non uniform heating of the membrane could also lead to a non uniform deformation of the membrane, which could decrease its mechanical properties.

1.6.5 Thermal uniformity measurements

Thermoreflectometry measurements were performed on membranes to analyze their thermal uniformity. The principle of this technique consists in measuring the relative variation of the light reflected by the membrane during heating (see Appendix C for more details). Since oxide did not have a high enough thermoreflectivity coefficient, we patterned polysilicon square pads uniformly distributed over the membrane to report the local thermal information. Pads were fairly spaced not to disturb the measurements but were not spaced too much to feature enough resolution. They were also designed fairly large in order to be easily detectable when the whole membrane was in the microscope

1.6. MICROHEATER CHARACTERIZATION AND RESULTS

field. Pads of 6x6 μm^2 spaced 21 μm from each other were consequently chosen (Fig. 1.21).

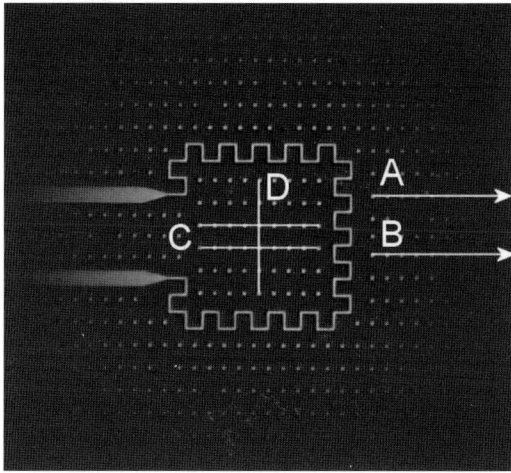

Figure 1.21: Picture of a test microheater on a 600x600 μm^2 membrane covered with polysilicon pads, during reflectometry measurements. The lines A, B, C and D show all investigated cut-views.

Measurements were performed on the accesses (Fig. 1.22(a)) and on one half of the membrane between the polysilicon heater and the substrate following cut views A and B (see Fig. 1.21 and 1.22(b)). The different membrane sizes were investigated and compared. Results depicted in Fig. 1.22 revealed that the thermal gradient was linear through the accesses but appeared slightly quadratic on the dielectric membrane side as expected by our simulations.

In the same way, a temperature diagram was extracted on the active zone delimited by the heater loop following cut views C and D (Fig. 1.21 and 1.23). These results revealed that both membrane sizes followed the same behavior with a non negligible temperature difference despite of the almost same power consumption. Along the direction perpendicular to the accesses (cut view D in Fig. 1.21), a maximum temperature difference of 50°C was observed. On the contrary, we observed a significant temperature decrease close to the accesses leading to a temperature difference of around 150°C between the accesses and their opposite. This decrease was not expected by our simulations. A hot spot zone appeared on the opposite side of the accesses due to the junction of the three

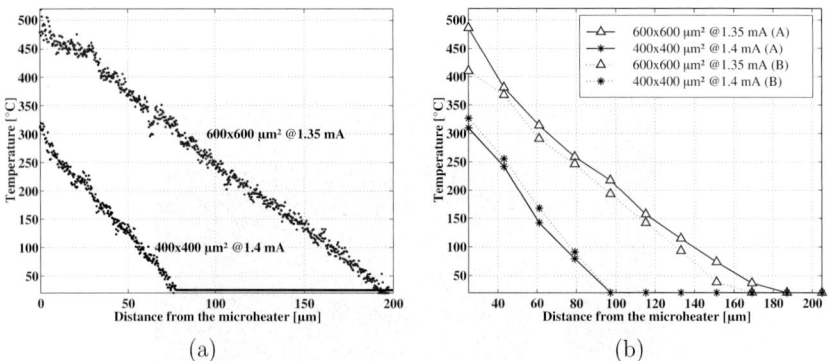

Figure 1.22: Results of the reflectometry measurements on the polysilicon accesses (a) and on the membrane between the heater and the substrate (b) for two membrane sizes.

arms of the loop. Same observations were made in [110] using an infrared camera. When the polysilicon was heated, its resistance increased more in this area and consumed more power. A solution would therefore be to decrease the resistance value of this arm of the loop by removing the small meanders and decreasing its section. In the same way, accesses could be drawn together to decrease their cooling effect. In spite of the increasing calculation time, simulation of the whole loop (with the accesses and meanders) could be required in this case to accurately improve the thermal study.

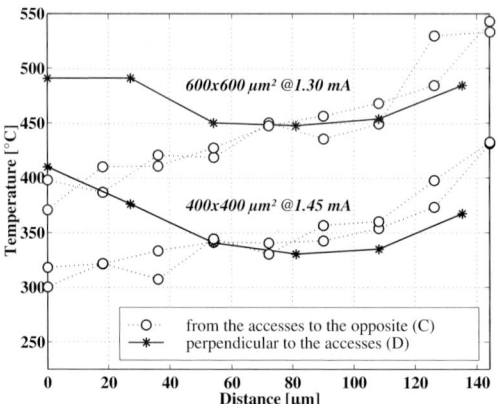

Figure 1.23: Results of the reflectometry measurements on the centered active area for two membrane sizes. Measurements were done following cut views C and D from Fig. 1.21.

1.6. MICROHEATER CHARACTERIZATION AND RESULTS

From the measurements on the 54 pads of the active zone of the membrane, a temperature diagram of this area was extracted by interpolation. Fig. 1.24(a) shows the results of this interpolation for a 600x600 μm^2 membrane. The picture confirms the hotter area on the opposite of the accesses as well as the cold points near the accesses.

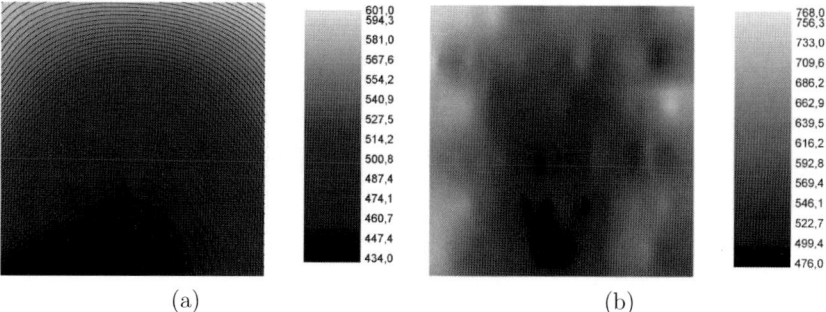

(a)   (b)

Figure 1.24: Temperature diagram on the active zone of a 600x600 μm^2 membrane heated at a current of about 1.4 mA. The accesses are on the bottom of the picture. (a) interpolated results from the 54 polysilicon pads covering the active zone, without metallic oxide; (b) reflectometry results from the interdigitated gold electrodes (4 fingers pairs) of an active zone covered with metal oxide.

Our simulations predicted that the metallic interdigitated electrodes covered by the metallic oxide would increase the thermal conductivity on the membrane and therefore decrease the thermal gradient on its center. This assumption was partly confirmed by our measurements. Reflectometry measurements were done without the help of the polysilicon pads distributed on the membrane but only on the gold interdigitated fingers. The extracted temperature diagram (without interpolation in this case) at the center of the active zone (delimited by the heater loop) can be seen in Fig. 1.24(b). The thermal uniformity was slightly increased and the cold points area appeared more confined.

In the literature, temperature diagrams are usually done on the polysilicon microheater only with an infrared camera, setting the emissivity to the emissivity of polysilicon [73][130]. We believe this technique is not so effective on oxide due to its very low emissivity. In our case, thanks to the additive polysilicon pads uniformly distributed on the membrane, we obtained very accurate temperature diagrams.

All our results showed that the thermal uniformity on the active zone of our membrane is not as good as expected by our simulations. Cold points were observed close to the accesses and hot spots on their opposite, featuring high thermal gradient in this direction. Perpendicularly to the accesses, temperature appeared more uniform, with fair gradient of about 50°C. Thermal uniformity could be increased by changing the position of the accesses as well as by decreasing the resistivity (and therefore the dissipated power) of the arm in the opposite direction of the accesses. On the other hand, our non-uniformity could be exploited to increase the selectivity to various gases by integrating electrodes on each region where the temperature is uniform.

Finally, our results clearly confirmed that wider membranes provide noticeably higher temperatures for a given power consumption.

1.6.6 Measurements at high temperatures

Experimental results

In previous subsections, all measurements were performed up to 400...500°C with a maximum current of 1.4...1.5 mA. Up to these values, the microheater heated with a quadratic resistance increase and cooled down to its initial room temperature resistance value, already from the first heating. In other words, its behavior was fully reversible. We will see now that if we increase more the current to reach higher temperatures, the resistance value at room temperature as well as its TCR will change.

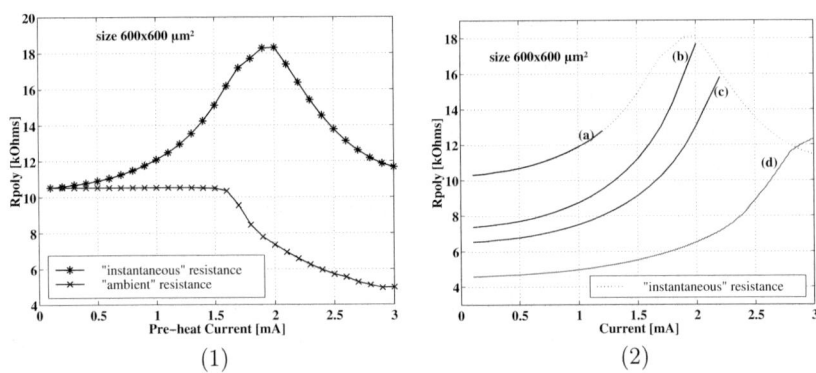

Figure 1.25: Measurements at high temperature for a 600x600 μm^2 membrane: (a) instantaneous and ambient resistance value versus pre-heat current and (b), current-resistance curves of pre-heated microheaters at 4 differents current values: (a) 1.2 mA; (b) 2 mA; (c) 2.2 mA and (d) 3 mA.

1.6. MICROHEATER CHARACTERIZATION AND RESULTS

Progressive heating cycles were performed on one 600x600 μm^2 microheater from 0.1 to 3 mA by increments of 0.1 mA, each followed by an instantaneous cooling. The resistance was measured during and after each of these 29 cycles as depicted in Fig. 1.25(1). The first "instantaneous" curve shows the resistor values measured during each heating while the "ambient" curve represents the resistances at the end of each cycle. The "instantaneous" curve is the same we would obtain by increasing the current until 3 mA for the first time on the polysilicon resistor. Its expected increase was followed by a slow decrease. During this heating, the resistor element was self-annealed and physically changes causing a resistance decrease as we will explain further. Upon rough removal of the current, the resistor quickly cooled to its "ambient" resistance value, i.e. the value measured after the polysilicon film had returned to room temperature (22°C). Until 1.5 mA, the ambient resistance kept the initial value but beyond this threshold value, the ambient resistance slowly decreased proportionally to the pre-heating current. It means that if we apply for the first time 2 mA on the microheater, for instance, after removal of the current, the "ambient" resistance will be irreversibly 30 % less than the initial resistor value.

Consider the case of a resistor pre-heated for the first time at 2.2 mA. When heating, resistance increased following the behavior of the "instantaneous" curve as depicted in Fig. 1.25. When cooling, resistance decreased to its "ambient" resistance value of 6.5 kΩ following the curve (c) in Fig. 1.25(2). When the same resistor was heated again until the same current value (2.2 mA), its starting resistance was 6.5 kΩ and it increased following the curve (c) of Fig. 1.25(2). Upon removal of the current, the resistance decreased to the same 6.5 kΩ value. The behavior of the resistor was reversible. Therefore, it appeared that the first heating changed the physical properties of polysilicon and fixed its ambient value for ever. From the second heating, the resistance value always followed the same curve. Note that an inflection point was observed on curve (d) in Fig. 1.25(2) revealing a higher resistance increase during heating from this point.

The behavior of the curves in Fig. 1.25(1) was confirmed by the literature [131][132] [133][123][134][135][109]. An inflection point in "instantaneous" curve occurs at the threshold current and corresponds to the onset of permanent resistance decrease. In the other hand, the maximum in the "instantaneous" resistance curve is reached at the second inflection point of the "ambient" curve.

The polysilicon resistor value decrease leads to a TCR change as mentioned in [136][131] [132][133][134][135]. To study the TCR evolution respectively to the pre-heat current, measurements were performed on 9 matched microheaters (on the same wafer), each

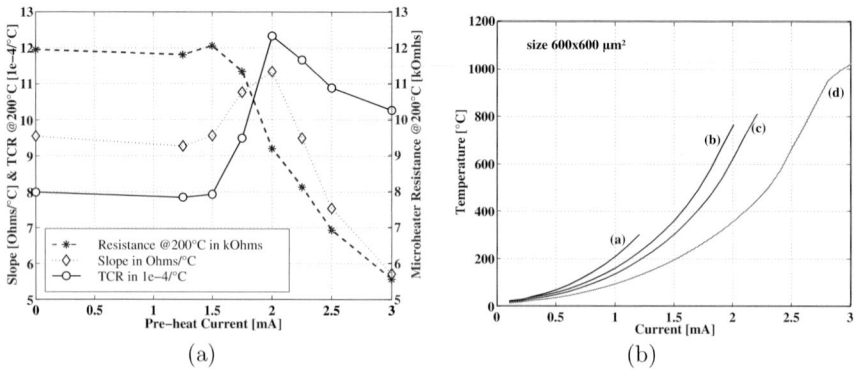

Figure 1.26: Measurements at high temperature for a 600x600 μm^2 membrane: (a) Microheater resistance at 200°C, slope and TCR at 200°C evolution versus pre-heat current measured on 8 microheaters pre-heated at differents currents (b), calculated current-temperature curves of pre-heated microheaters at 4 different current values: (a) 1.2 mA; (b) 2 mA; (c) 2.2 mA and (d) 3 mA.

being pre-heated at an incrementally increasing current of 0.5 mA. As for previous calibrations, measurements were taken at 10 μA over a temperature range from 22°C to 250°C in 50°C increments. Fig. 1.26(a) shows the measured resistance at 200°C, the slope of the resistance-temperature curve and the calculated TCR at this temperature versus the pre-heating current. As explained above, the TCR was calculated as the ratio between the slope and the resistance at 200°C. Its value did not change below the threshold current (1.5 mA), as it would be expected since the slope and the resistance at 200°C were constant on this range. From this value, TCR increased until 2 mA and drastically decreased afterwards. Until 2 mA, the slope increased when the resistance decreased, leading to a TCR increase. After 2 mA, the slope decreased in the same time than the resistance and therefore leads to a slight TCR decrease. This curve appeared to follow the "instantaneous" resistance value, i.e. its maximum occurring at the second inflection point of "ambient" curve (Fig. 1.25(1)). This result can not be compared with literature since TCR there appears sometimes increasing, sometimes decreasing [136][131][132][133] but seems absolutely coherent in our case.

From the TCR results, the current-temperature curves were calculated for each points of Fig. 1.25(2) to extract the curves of Fig. 1.26(b). It demonstrated that approximately 1000°C are reached at 3 mA without membrane or aluminum contacts destruction. It is also interesting to fit the threshold current value with the temperature. Polysilicon is known to recrystallize at temperatures around 600°C [123][109][137], just below its tem-

1.6. MICROHEATER CHARACTERIZATION AND RESULTS

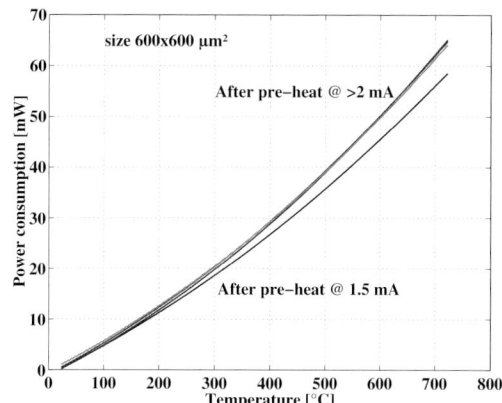

Figure 1.27: Power consumption versus target temperature for microheaters pre-heated at different temperatures (membrane size 600x600 μm^2).

perature deposition. By extrapolating the curve (a) in Fig. 1.26(b), the recrystallization temperature appeared to be reached at about 1.5 mA, i.e. exactly at the threshold current value from which the ambient resistance started to decrease slowly.

Finally, the power consumption versus temperature was calculated and displayed in Fig. 1.27 for 4 resistors pre-heated at 4 different temperatures. A significative difference was observed during pre-heating from 1.5 to 2 mA while TCR and instantaneous resistance increased. The TCR increase improved the sensitivity to temperature measurements but increased also the power consumption since resistance increased faster. An increase of around 5 mW was observed at 700°C. Beyond 2 mA, the TCR decreasing in same time than the "ambient" resistance did not affect the power consumption anymore.

Additional observations were reported in [109][137][108]. They observed that the ambient resistance value after heating depends not only on its thermal history (as observed here) but also on its cooling rate. Furthermore, heating rate was reported not to have any impact on the resistance value [109]. They reported that after heating at around 900°C, a cooling rate of 0.02 K/s can increase the ambient resistance value to 33 % of its initial value. On the contrary, an ambient resistance value decrease by 17 % can be achieved with cooling rate of 12.1 K/s. Tests in [109][137][108] were performed on lower doped polysilicon resistors in vacuum.

Our measurements were always performed with instantaneous cooling rate and gave resistance decrease of more than -40 % after heating at ∼900°C or ∼2.3 mA. In sensors applications, it can not be considered to achieve cooling during a long time. Nevertheless it could be envisaged to slowly cool the membrane after the pre-bake of our gas sensitive layers at 700°C if it could lead to a noticeable performance increase. We reported a slight power consumption increase after pre-heating at 700°C due to the resistance and TCR changes. When decreasing the cooling rate, we could avoid a resistance decrease (as observed in [109]) but we can not expect that the TCR would stay constant. This point is not discussed in [109].

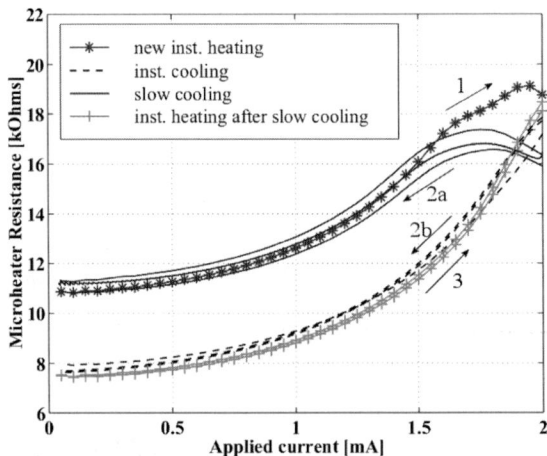

Figure 1.28: Polysilicon resistance versus applied current for different heating-cooling cycles.

Some tests were performed with lower cooling rate as a verification. Cooling rate of 50 μA/min (or 17.5°C/min) was chosen after heating up to 700°C (2 mA), i.e. during 40 minutes. The graph in Fig. 1.28 depicts the results comparing 3 tests at instantaneous cooling with 3 other ones at slow cooling rate. A new instantaneous heating was firstly done on one microheater (1). Heating was instantaneously performed since it does not have any impact on the polysilicon resistor. This heating was followed by a slow cooling with a rate of 50 μA/min on a first microheater (2a). On a second one, the cooling was done instantaneously (2b). A room temperature resistance drift was observed between both cooling rates. In the first case, the room temperature resistance was observed unchanged but on the second one, it appeared reduced by 30 %. Note that it was also

1.6. MICROHEATER CHARACTERIZATION AND RESULTS

observed at the beginning of the slow cooling (at 2 mA) that the resistance strongly decreased before its normal decrease. This behavior will be explained further in ageing tests. Finally, a heating was done on both cooled heating resistors (3) and we observed strangely that heating curves started from the same room temperature resistance for both microheaters. It seemed that at the end of the slow cooling, when the current was switched off, the room temperature resistance was instantaneously decreased. In conclusion, we observed that slow cooling rate showed a different behavior than an instantaneous one but in fine gave the same room temperature resistance. Therefore, our measurements differed from ones in [109][137][108] on this point.

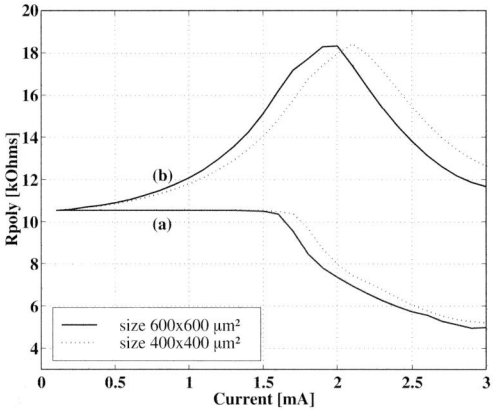

Figure 1.29: Comparison of instantaneous and ambient resistance value versus pre-heat current for 2 membrane sizes.

All of these discussions were carried out for 600x600 μm^2 membranes. A comparison was performed with 400x400 μm^2 membranes and results are depicted in Fig. 1.29. The graphics reveal that the threshold current as well as the maximum instantaneous resistance value were shifted to the right of around 0.15 mA. As the substrate is closer to the heater in 400x400 μm^2 membranes, more power is needed to reach a given temperature. Therefore, more power is also needed to reach the irreversible physical effect in polysilicon.

It could be interesting to analyze the correlation between heating and current effect on the resistance decrease. It is impossible to put the sensor in a oven at more than 500°C to verify if the current has an effect or not on the polysilicon physical change. But other analyses were performed. Our previous results showed a shift of the threshold current to the right when we decreased the membrane area. In both cases, the microheater is identical as well as the resistance value at ambient temperature. We know that the temperature we reached for a given current was different in both cases as depicted in Fig. 1.16. Therefore the shift between the threshold currents demonstrate that the resistance decrease is dominated by the temperature rather than the current. To confirm this assumption, we applied the same current ramp on an identical microheater on substrate. The large heat capacity of the substrate avoided the microheater to heat until the temperature reached on membrane for the same current. We did not observe any difference between the first and the second heating, confirming that the microheater was perfectly reversible (with a linear behavior) up to 3 mA when it was built on substrate. The same observation was done when reading the literature. The threshold current measured in [136][131][132][133] was always higher than our measured currents since their polysilicon resistors were put on substrate which prevented sufficient heating.

Thanks to the great isolation of our polysilicon microheater, we observed under microscope visible light emission from the polysilicon loop when current value reached 1.9-2 mA, i.e. at temperature close to 700°C or when the "instantaneous" resistance value reached its maximum. This behavior was already observed in [102] in vacuum when temperature reached 630°C. Miniaturized low power IR sources are fabricated in this manner using micromachining to isolate a polysilicon filament [102]. These devices are to be used as the calibration source for space detectors. A light emitting region of 30x40 μm^2 is achieved and needs for around 2 mW to reach 630°C. Works of [129] on porous silicon membrane reported a visible light emission at 550°C.

It must be noted that this pre-heating of polysilicon resistors is commonly used to adjust their values separately after IC fabrication and packaging to compensate the process deviations (such as in polysilicon sheet resistance) [136][131][132][133][134][135]. In this case, the pre-heating is called "trimming" and the current is applied in controlled pulses for a more precise control and to avoid rise of temperature in the resistor periphery [135]. Furthermore, to achieve more precise resistors targets, a portion of the trimmed resistance can be recovered to a higher value of resistance by re-applying a current of amplitude less than the final trim current but above the thresold current. The recovery is dependent upon the severity of the resistor trim and upon the recovery time [133]. The magnitude of recovered resistance is very small in comparison to the amount which is trimmed but usable. Since it is repeatable, it offers an interesting method to finely tune resistors which are trimmed past their target. But this technique cannot be applied in our case.

1.6. MICROHEATER CHARACTERIZATION AND RESULTS

Physical and structural behavior

Polysilicon is composed of crystalline grains that are disoriented relative to one another and joined at defect laden grain boundaries [133]. The room temperature resistance reduction is based upon physical and structural changes at the highly resistive grain boundaries when large currents are flowed into the polysilicon resistor. On the contrary, the grain resistance which constitutes the lower part of the film total resistance, remains unchanged [136]. Two mechanisms of resistance reduction are proposed and detailed in [131], [133] and [136]. First, structural changes occur at the grain boundary, due to the extreme heat and high fields generated during current increase, resulting in a more coherent transition (or melting) between grains. This portion of resistance reduction results in a permanent resistance change [133]. Second, the dopant concentration locally increases within the grain boundary by "liquid phase" segregation [133]. Once sufficient Joule heating is reached, melting occurs radially about the points of grain-to-grain contact. During cooling, after the current is stopped, the melted grain boundary region cools and radially solidifies towards the center of the grain boundary. Since dopant atoms have an affinity for the liquid phase, they migrate from the solid phase region into the still liquid phase region, in the center of the melt, and decrease the grain boundary resistance. This dopant segregation provides a smaller, reversible component to the resistance change. It is interesting to note that [135][132] reported that the polysilicon grain size does not increase during the heating process.

The resistance recovery occurs when the localized concentration of dopants, which segregated to the melted grain boundary during pre-heating, break free and re-distribute into the grains through a diffusion process [131]. As the current applied is less than the pre-heating current, the heat generated at the grain boundary is insufficient to cause localized melting to occur [133]. However, this heat provides sufficient energy to allow redistribution of dopants which results in an increase of the polysilicon resistance. For more details about this behavior, see [131][132][133].

To explain the TCR change, a model is developed in [136] and demonstrates that the TCR after pre-heating can increase or decrease with decreasing of the grain boundary resistance depending on the kind and doping level of the impurities in the resistor. Their model is not able to explain the behavior observed in our case, the TCR increasing followed by its decreasing. Nevertheless, it seems coherent that a redistribution of dopants within the grain boundary region results in a non-monotonic TCR evolution.

Summary

If it is needed to increase the membrane temperature until 700°C, it must be known that the room resistance as well as the polysilicon TCR will change. Therefore, the polysilicon resistor calibration must be achieved after a pre-heating at the working temperature. We observed that a pre-heating of our polysilicon resistor around 700°C (i.e. 2 mA) increased the TCR by 0.018 % and decreased its resistance by 30 %. The TCR increase improves the sensitivity to the temperature measurements but increases the power consumption of around 5 mW at 700°C. Finally, we measured that a power consumption of around 60 mW is sufficient to perform the annealing of the gas sensor active layer at 700°C.

1.6.7 Thermal ageing

Previously, all measurements were performed instantaneously, during short period of time. We observed the behavior of the resistance when heating or cooling as well as after heating-cooling cycles. We did not observe the resistance evolution at a given temperature along the time. So, to complete our study, thermal ageing tests were performed on microheaters to control that they could maintain their functionality during a long period of time. Some devices were maintained to typical operating temperatures during a long period of time and the resistance value of the heater was taken as the parameter to monitor during our tests.

As it will be discussed, we observed a different evolution of the microheater resistance and temperature depending on the power supply. So, in the following we will separate the tests performed at constant current from those performed at constant voltage.

Tests at constant current

Firstly, microheaters were heated during several hours at a constant current of 1.4 mA (i.e. 400°C, the maximum gas sensor working temperature) as well as at 1.5 mA (the threshold current before the irreversible regime) on two sensors with different history. Results of normalized resistance evolution versus time are depicted in Fig. 1.30. Polysilicon resistor value was observed to slightly increase during the heating with a slope depending strongly on the current value and the heating history. A variation of up to 12 % has been observed after 15 hours at 1.4 mA and a higher one of up to 39 % after heating at 1.5 mA. The third curve depicts a higher variation in spite of the same current value due to the different thermal history. A resistance recovery appeared here as it will be explained further.

1.6. MICROHEATER CHARACTERIZATION AND RESULTS

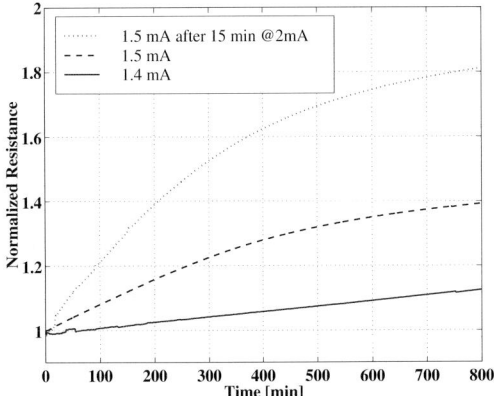

Figure 1.30: Behavior of the heater resistance versus time depending on the current applied and the thermal history.

So the resistance increased according to stress time, depending on the temperature applied and the thermal history of the resistor. When constant current was applied, we measured a voltage increase. Consequently, the power consumption UI or RI^2 was draised, leading to a temperature increase. With the temperature increase, the resistance increased more since its increase is a function of the temperature. The resistance drift of 12 % during heating at 1.4 mA for 15 hours leads to an unacceptable temperature increase of about 110°C. This temperature increase was confirmed by measurements on flow sensors. During the heating, the voltage increased across the thermopiles such as it would be the case if temperature increased.

Since temperature increased with the resistance increase, when heating at 1.5 mA, we passed through the threshold temperature and entered into the irreversible domain. To confirm this assumption, the TCR was measured before and after the heating in order to explain the resistance increase. All our results are summarized in the calibration curves in Fig. 1.31. The TCR was observed unchanged after the heating at 1.4 mA since the increasing temperature was not high enough to exceed the threshold value. On the contrary, it was increased by about 29 % after the heating at 1.5 mA, passing from 8.16 to $10.53 \times 10^{-4} °C^{-1}$. When comparing this result with those displayed in Fig. 1.26, the same TCR increase was observed when increasing the pre-heating current up to 1.8 mA which confirms that the irreversible domain was reached.

The room temperature resistance was increased by about 10 % after this heating test

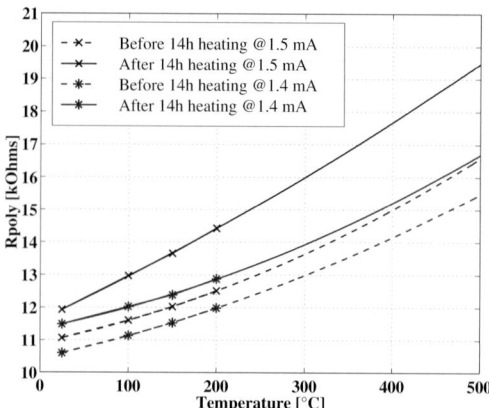

Figure 1.31: Polysilicon resistance calibration before and after heating at 2 different temperatures.

in both cases. In fact, the room temperature resistance actually increased by about 10 % when heating at 1.4 mA. But when heating at 1.5 mA, the room temperature resistance increase was countered by a slow resistance decrease when temperature exceeded the threshold value. Fig. 1.25 reported a room temperature resistance decrease by 20 % at the end. So an actual resistance variation of about 20 to 30 % seems more consistent in this case.

As observed in Fig. 1.30, we analyzed also the resistance relaxation when the microheater was previously heated at 2 mA. During this pre-heating, the resistance at 2 mA appeared not to change significantly. At the end of these 15 min at 2 mA, the extracted TCR was increased and the room temperature resistance, decreased by about 30 %. When following this pre-heating by a heating for 15 hours at 1.5 mA, we observed that the room temperature resistance was recovered. The room temperature resistance after heating was increased by about 7 % in comparison with the original resistor value (before the pre-heating at 2 mA); leading to a total variation of about 37 %. Furthermore, the resistance value at 1.5 mA at the end of the heating was measured to be almost the same as the resistance at 1.5 mA without pre-heating. We demonstrated here the case of a complete resistance recovery when heating for long time at a current lower than the pre-heating current. This point was never mentioned in literature.

A detailed study on the first 15 min. of heating was also achieved and revealed that the polysilicon resistance started to roughly decrease before its slow increase (Fig. 1.32).

1.6. MICROHEATER CHARACTERIZATION AND RESULTS

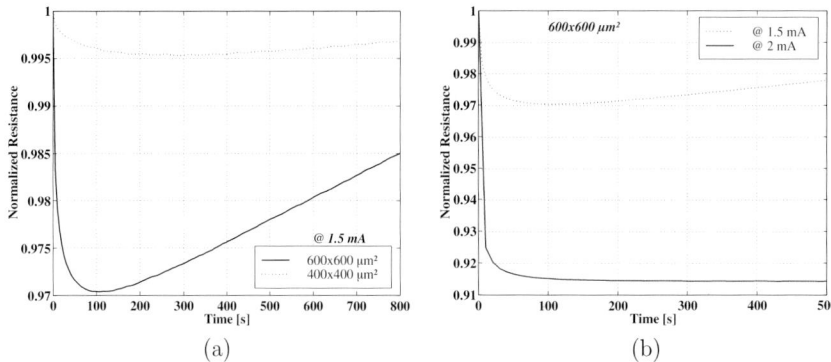

Figure 1.32: Microheater resistance versus time on 2 different membrane sizes and at 2 different applied currents; (a) comparison of both membrane sizes at 1.5 mA and (b), comparison of 2 different heating currents on a 600x600 μm^2 membrane.

This behavior was observed to be reproducible on a lot of identical measurements. The relaxation of the polysilicon resistance during the first minutes was reported in [109] and [137] as observed here. The resistance decrease was observed proportional to the applied current and the reached temperature, such as the following increasing slope. In this way, the difference of slope between the two membrane sizes we can observe in Fig. 1.32(a) was not due to the difference in membrane areas but to the different temperatures reached on both membrane sizes when same current was applied. Similarly, Fig. 1.32(b) displays the quick resistance decrease before the slow resistance increase when heating at 1.5 mA and the constant evolution when heating at 2 mA.

Finally, power consumption versus temperature was compared before and after a heating during 15 hours. It was observed slightly increased by an amount proportional to the room temperature resistance increase, i.e. around 2.5 mW at 400°C even when TCR increased after heating at 1.5 mA.

Tests at constant voltage

When applying a constant current on the microheater, the temperature increased with the heated resistance increase. Another solution consists in applying a constant voltage instead of a constant current along the time. In this case, the resistance increase at high temperature leads to a current decrease. Consequently, the power consumption equal to RI^2 globally decreases since R and I respectively increases and decreases proportionnaly but I appears squared. Therefore, the temperature decreases with the decreasing power consumption and partially offsets the resistance increase with the temperature. This behavior was confirmed by comparing measurements at constant current and measurements at constant voltage we can see in Fig. 1.33. Both tests were achieved starting with the same power consumption and therefore the same temperature.

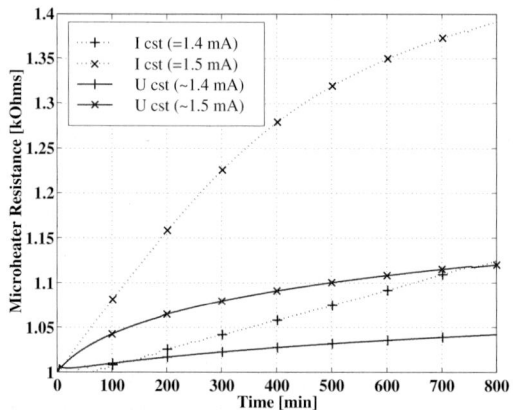

Figure 1.33: High temperature resistance increase versus time. Comparison between constant current and constant voltage applied, both providing the same power consumption and therefore the same temperature at the beginning of the test.

According to Fig. 1.33, the resistance increase clearly appears slower at constant voltage than at constant current. In both cases the resistance relaxation at high temperature appears as a resistance increase. Nevertheless, at constant voltage, the power consumption decrease leads to slow down its increase while at constant current, the power consumption increase leads to speed up its increase. An advantage of the constant voltage supply is that if the voltage applied is fixed below the threshold of irreversibility, the TCR will never change thanks to the temperature decrease.

The temperature decrease in this case was confirmed by flow sensors measurements as a function of time. The voltage measured across the thermopiles decreased when constant voltage was applied on the microheater. Our measurements clearly confirmed that the temperature decrease at constant voltage versus time was slower than its increase at constant current.

The temperature decrease was also confirmed by reflectrometric measurements. It seemed interesting to extract the temperature independently of any polysilicon resistance.

1.6. MICROHEATER CHARACTERIZATION AND RESULTS

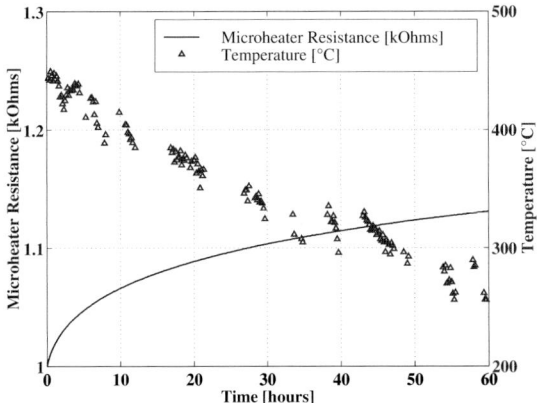

Figure 1.34: Temperature decrease versus time at constant voltage applied on the microheater.

Using the reflectometry measurements as external temperature sensor, we extracted the graph shown in Fig 1.34 when heating a microheater at constant voltage during 60 hours. It reveals an unacceptable temperature decrease of about 180°C during 60 hours of operation. Nevertheless, this result needs perhaps to be interpreted with care since other factors in the measurement procedure could also contribute to increase the relaxation behavior.

At first sight, heating at constant voltage appears more interesting since the temperature decreasing is slower than the temperature increase when constant current is applied. Furthermore, when heating at high temperatures up to the threshold value, TCR will always remain constant at constant voltage, which is not the case at constant current.

Similar resistance increase over time, at constant voltage, was reported in [105][97] and [110] along 10 weeks with a logarithmic drift and a slope depending on the current applied. Variations up to 15 % were reported, depending on the temperature applied. This lack of temperature stability versus time is unacceptable and needs to be solved. An idea proposed in [105][97][110] is to perform a stabilization step at the operating temperature during about a week before using the microheaters for gas sensors in order to assure a temperature stable during the operation time of the sensor. Nevertheless, if this operation is done at constant voltage, the starting temperature of the pre-ageing step needs to be fixed at a higher value than the operating temperature to take into account its decrease with time.

We proposed another solution which consists in monitoring the voltage as well as the current across the microheater during heating in order to maintain constant the power injected into the microheater. By this way, the temperature would stay constant even if resistance would change. Thanks to the ability of our microheater to be fully integrated with close electronics, a dedicated circuit could achieve this operation. The recent works of [95] reported an interesting example of on-chip temperature controller to regulate the temperature of the membrane but did not provide a compensation of the polysilicon ageing effects. Another interesting solution and easy to perform, but at the detriment of the power consumption, would be to connect another resistor (featuring the same value) on the silicon bulk in series to the microheater. The same current would flow through the two polysilicon resistors, heating the one on the membrane without changing the additional resistor value on the substrate. If we supply the two resistors at constant voltage, the additional resistor would fix the power in the microheater during its relaxation: if the microheater resistance increases, the current flowing through it would decrease, leading to an increasing voltage through the microheater.

Physical explanations

After short heating-cooling cycles up to very high temperatures, we observed that the room temperature resistance followed the behavior shown in Fig. 1.25(1) depending on the applied current. The resistance firstly stayed constant before a slight decrease after the threshold current. We gave a physical explanation for this decrease here above.

On the other hand, we observed a different evolution of the room temperature resistance when maintaining the polysilicon resistor at high temperature for long time. The room temperature resistance slowly increased until the threshold current was reached and decreased after this value. Furthermore, it was observed that the room temperature resistance was recovered if a heating at a current lower than the threshold value was performed for long time after a heating at a current higher than the threshold value. We finally observed a quick room temperature resistance decrease during the first minutes of heating before its slow increase.

These observed behaviors are confirmed in the literature. [111] (or [138]) reported a constant room temperature evolution when heating during 4 hours at 400°C. This constant behavior was followed by a slight increase when heating at temperatures between 400 and 600°C. Finally, after 4 hours heating at temperatures higher than 800°C, the room

1.6. MICROHEATER CHARACTERIZATION AND RESULTS

temperature resistance roughly decreased. The results reported in [111] revealed also the resistance recovery such as observed in our case.

Therefore, our previous physical explanations need to be completed. The room temperature increase when heating for long time can be explained by changes in the concentration of electrically active dopant atoms [111]. When quick heating-cooling cycles are achieved below the threshold, the room temperature resistance does not have the time to change. On the contrary, during long heating, phosphorus atoms slowly diffuse towards the grain boundaries and become electrically inactive [111]. At higher temperatures, the increase in resistance due to this diffusion of atoms is counterbalanced by the dominating influence of the structural change at the grains boundaries and the liquid phase segregation (see above), causing the resistance to decrease. On the other hand, as reported above, the observed resistance recovery was observed here thanks to the long time required for the dopants to redistribute in the resistor. Finally, the quick room temperature resistance decrease during the first minutes of heating could be explained by the same theory of structural change and segregation reported above but appearing reduced in this case.

Summary

When maintaining the polysilicon microheater at operating temperature for long time, we observed during heating, a strong resistance increase leading to an unacceptable temperature drift with time. If a constant current supplies the microheater, the power consumption increases over time, leading to a temperature increase. On the contrary, if a constant voltage is applied on the microheater, the power consumption decreases as well as the temperature.

In the same way, the room temperature resistance was affected by long heating operations. We reported its slight increase up to the threshold temperature. We also demonstrated how the resistor was recovered when previously heated at a temperature higher than the threshold temperature. We gave physical explanations to our observations in order to complete those reported in the previous section.

To conclude, we recommended to perform pre-ageing tests before using the microheater for long time in order to assure a temperature stable during the operation time of the sensor. We also proposed a more accurate technique which consists in maintaining the injected power constant during heating in order to stabilize the temperature when the polysilicon heater resistance increases. This operation could be achieved by a dedicated circuit integrated close to the sensor.

1.6.8 Reliability

Microheaters on membrane are fragile devices as they are based on very thin and large dielectric membranes. It is necessary to ensure good yield of the structure, both at the fabrication level and also during the operating conditions [110]. We have seen how the backside etch and chip dicing and packaging were considered the most critical processes concerning mechanical yield. After packaging, our sensors exhibited high mechanical robustness as shown during the numerous shippings from one lab to another, from one measurement setting to another.

A deep study was performed about the ageing of the microheaters and slight resistance variations were observed versus time. We concluded that the power consumption and not only the current or the voltage applied through the heating resistor needs to be monitored over time to maintain the temperature constant when resistance changed.

Interferometric measurements were carried out on membrane to analyze their deformation under heating. Deflections were measured when increasing the current applied on the microheater until the membrane broke down. Measurements revealed that the membrane deformation was negligible up to 1.5 mA. Nevertheless, when increasing the current beyond this threshold power, the deformation of the heated area increased strongly and not uniformly before breaking. Same observations were reported in [116]. Deflections higher than 2.5 μm were observed on some hot points of the microheater at temperatures around 1000°C (3 mA). Thanks to the high robustness and elasticity of the membrane, it did not brake for currents values up to 4 mA.

Thermal shocks were finally performed as proposed in [110], [99] and [98]. Switching between 0 and 1.5 mA (corresponding to abrupt temperature variations from 0 to $\sim$400°C and reversely) were carried out at $\sim$3 Hz during 15 minutes on some microheaters lying on 600x600 μm^2 membranes. During these tests, resistance variations were monitored as an indicator of the ageing. Same resistance drifts were observed, starting from a rough decrease and followed by a slight increase. Drifts appeared more slowly than when constant current or voltage was applied on the resistance but were a little bit more severe. A drift of about 3 % was measured on the room temperature resistance at the end of the test.

Similarly, harsher tests were achieved by switching microheaters on 800x800 μm^2 membrane from 0 to 3 mA and reversely (corresponding to abrupt temperature variations between 0 and $\sim$1000°C) at the same frequency of $\sim$3 Hz during more than 5 minutes to study its mechanical strength. Alternating membrane deflections as well as flashing light emissions were clearly visible under the microscope during this test. At the end of these tests on 2 sensors, we did not observe any membrane damage which confirms the mechanical and thermal robustness of our membranes.

1.7 Conclusions

A new polysilicon loop shape microhotplate was presented. Its fabrication allowed it to be fully compatible with a standard SOI-CMOS process in order to built a complete smart gas sensor, integrating on the same chip our microhotplate, the additional steps for gas sensing and the close electronics for signal treatment. Chapters 3 and 4 of this part of the book will study the additional steps to complete our microhotplate in gas sensor and will validate the complete CMOS-SOI compatibility of our design.

The microhotplate was designed to be supported by a 1 μm thin stacked dielectric membrane, such as studied in the Chapter 2 of the first part, in order to reduce the power consumption. We showed that the membrane temperature over consumed power ratio of our microhotplate polysilicon heater was found to be 15°C/mW which is an excellent result in comparison with recent published microheaters results. This value was confirmed by our measurements as well as by our thermal simulations. The high strength of our membrane was also demonstrated during our tests of realibility and confirmed the calculations of residual average stress and strain gradient previously presented in the second chapter of the first part.

The membrane was released in post-processing using our optimized fully compatible TMAH-based etching technique. We demonstrated this technique to be powerful in our case to etch silicon while the whole wafer was immersed in the etchant during 3 hours despite of the presence of aluminum on the both sides of the wafer. We also concluded that this step was very easy to perform on SOI wafers thanks to the buried oxide layer used as etch stop.

The loop shape of the microheater was designed using ANSYS simulator in order to reach a high thermal uniformity at the center of the membrane. The design was optimized to avoid the use of additional aluminum, polysilicon or silicon spreaders which increase the complexity of most recent designs. Simulations revealed great thermal uniformity (as well as a very high thermal insulation) with variations of maximum 50°C on the whole 240x240 μm^2 active heated area, without the oxide metallic layer on top. Our thermoreflectometry measurements approximately revealed the same variations in the direction perpendicular to the accesses. Unfortunately, we found that the accesses introduce cold points leading to high thermal gradients from the accesses to their opposite position. We proposed a solution for a future design in order to improve the thermal uniformity of the heated area or to exploit the non-uniformity by integrating several electrodes to improve the selectivity to gases.

Our measurements confirmed the two main drawbacks of the polysilicon, when it is used as microheater: i.e. its change of resistivity (and TCR) at high temperature, altering previous calibrations, as well as its poor long-term thermal stability. Temperature as high as 700-800°C are needed to anneal gas sensitive layers for their better stability at the operating temperature. A deep study was done in order to understand all of these behaviors and to be able to foresee the characteristics of the polysilicon resistor at high temperature as well as during time. This study completed by some physical explanations provides an interesting contribution to the literature, incomplete on this subject, as our reviews revealed. We observed that heating polysilicon at temperature around 700-800°C decreased its room temperature resistance and increased its TCR. Our ageing tests revealed that the room temperature resistance was fully recovered after heating at the operating temperature during 1 night, demonstrating its reversibility. Our ageing tests also showed that pre-ageing tests are essential before using the microhotplate for long time in order to assure a temperature stable during the operation of the sensor. We proposed another technique which consists in monitoring the injected power during heating instead of the only current or voltage in order to stabilize the temperature.

The calibrations of our microhotplate before their measurements allowed us to measure the average temperature on the membrane during heating. Calibrations of a lot of representative microhotplates (with the same design) dispersed on various wafers and at different locations of the same wafer, allowed us to extrapolate our calibration results to any microhotplate with the same design, preventing a tedious calibration of each microhotplate separately. Finally, our measurements up to the standard operating temperature of gas sensors revealed some interesting conclusions and explanations especially about the geometry of the loop and the membrane size.

1.7. CONCLUSIONS

Unfortunately, our design revealed two main drawbacks. We speaked about the disappointing thermal uniformity. We must still note the high bias needed to reach the operating temperature, i.e. approximately 20 V to reach 400°C. This high bias, even when the device is supplied at constant current, complicates the co-integration of our microhotplate with low power circuits. Furthermore, to allow the modulation of the operating temperature (as required in some cases to improve the selectivity to gases), we need to reach quickly and easily the operating voltage, which is delicate in this case. Nevertheless, a new design could be conceivable by separating the microheater in 2 resistors in parallel, dividing the voltage by 2 and with twice higher current passing through the microheater. With two opposite accesses, this new design could also improve the thermal uniformity over the heated area.

The following chapter integrating our microhotplate as well as a new heating resistor wire for flow sensing applications will complete our study. Results on the response time of our microheater and on the thermal gradients provided by our membrane will be presented.

Chapter 2

Microheater based flow sensor

2.1 Introduction

A successful microheater constitutes a very attractive basic clock for building a flow-sensor for which a growing demand exists in industrial to medical applications. Many different sensing principles (such as thermopiles, diodes, microbridges, ...[139]) are used and impose a complicated trade-off between low power consumption, high flow velocity range, sufficient flow sensitivity and compatibility with IC processes. The challenge is to carefully take all of these parameters into account for designing a new high performance flow sensor.

This chapter reports an improved low cost directional flow sensor (Fig. 2.1), fully compatible with IC processes, with a fair sensitivity on a large flow velocity range (from 0 to 8 m/s) and a short response time at a very low consumption (30 mW). Our sensor outperforms recent realizations since it presents attractive trade-offs between these parameters for a large range of applications.

The flow sensor presented here is of the calorimetric type, based on the differential temperature measurement of a directional flow upstream and downstream of a heater. When there is no flow stream, the temperature profile is symmetrical around the heater. The symmetry is disturbed when a gas flows, since the upstream detector is cooled by the oncoming gas, while the gas is heated by the hot heater and then, in turn, flows heat to the downstream detector [13]. Thus a temperature difference between the cooled upstream area and the less cooled (or heated) downstream area occurs. This temperature difference can be converted into an output voltage, which is used as a measure for the fluid velocity and flow direction [140]. The advantage of this differential measurement method is the

Figure 2.1: Picture of a 800x800 μm^2 flow sensor with one direction of detection.

elimination of offsets due to factors such as ambient pressure and temperature.

First of all, the designs we developed are discussed. Since the fabrication of our flow sensors exactly followed the microheater fabrication, this will not be detailed again. Electrical measurements without flow are then explained and will reveal some interesting observations about the microheater, such as his response time and the thermal gradients on the heated membrane. Finally, the results of our flow measurements are described and compared with the state-of-the-art in the literature.

2.2 Design and fabrication

For initial simplicity, only one direction of detection was firstly integrated[1] (from left to right or reverse as shown in Fig. 2.1). A microheater was used to heat our 1 μm thick dielectric membrane (3 dimensions were tested: 800x800, 600x600 and 400x400 μm^2 depending on the configuration). Thermopiles or pre-calibrated polysilicon resistors were used to measure temperatures upstream and downstream of the heater.

---

[1]Sensors combining flow sensitivity in two directions perpendicular to each other were designed but not measured at this time. They would allow to determine both the velocity and any direction of wind over the surface of the chip.

2.2. DESIGN AND FABRICATION

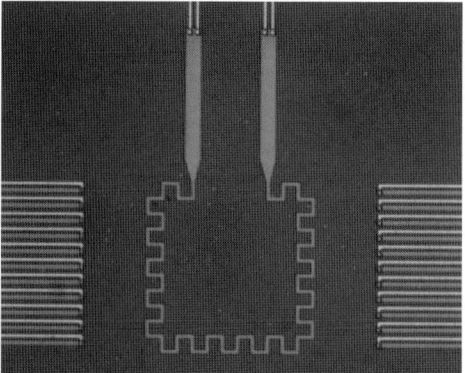

Figure 2.2: Close-up view of the flow sensor in configuration 1. The microheater and the thermopiles hot points can be seen.

Thermopiles are two-lead elements that measure the temperature difference between the ends of the wires. They are based on the thermoelectric Seebeck effect: a temperature difference ΔT in a conductor creates an electrical voltage ΔV [13],

$$\Delta V = \alpha_S \Delta T \qquad (2.1)$$

where α_S is the Seebeck coefficient expressed in V/K. It is a material constant. By taking wires of materials with different α_S (α_{Sa} and α_{Sb}), with a junction of the wires at the hot point (at temperature $(T + \Delta T)$), and their other tips at the cold point (at temperature T), the voltages measured across the two wires (ΔV_a and ΔV_b) are subtracted and an effective Seebeck coefficient α_{Seff} remains [13]:

$$\Delta V = \Delta V_b - \Delta V_a = (\alpha_{Sb} - \alpha_{Sa})\Delta T = \alpha_{Seff}\Delta T \qquad (2.2)$$

To obtain a thermopile with a fair effective Seebeck coefficient, we must choose a combination of a first material with a high coefficient (α_{Sb}) and a second with a negligible or a negative one (α_{Sa}). Connecting N junctions (or strips) in series allows to obtain a resulting thermopile with a higher sensitivity S_{th} [13]:

$$S_{th} = N\alpha_{Seff} = N(\alpha_{Sb} - \alpha_{Sa}) \qquad (2.3)$$

Thermopiles used here were polysilicon/aluminum couples, polysilicon having a doping level fixed by the transistor gates doping leading to a sheet resistance equal to 27 $\Omega/\square$. Usually, polysilicon used in thermopiles needs to be less doped to increase its Seebeck coefficients but the value of 27 $\Omega/\square$ was imposed by our standard IC process. For comparison, [13][141] reported typical Seebeck coefficient on the order of 200 to 500 $\mu V/K$ for 0.4 μm thick, 250 to 2500 $\Omega/\square$ phosphorus doped polysilicon film (332 $\mu V/K$ found in [142]). The Seebeck coefficient of our highly doped polysilicon is unknown but can be expected quite low. It will be approximated by temperature measurements on the membrane using pre-calibrated polysilicon resistors. On the contrary, the Seebeck coefficient of aluminum is known to be very small (about -1.7...-3.2 $\mu V/K$ [13][141]) and therefore, the influence of aluminum lines and aluminum interconnections are usually not significant. Nevertheless, [143] reported that aluminum/polysilicon thermopiles suffer from heat conductance to the substrate through the aluminum lines. [144] reported coefficients of 58 $\mu V/K$ for an aluminum/hightly-p-doped thermopile. As a comparison, [145] reported the highest sensitive thermopile built in a standard CMOS process we can find in literature, using a n-poly/p^+-silicon, showing Seebeck coefficients of respectively -320 and 430 $\mu V/K$, resulting in a total Seebeck coefficient of about 750 $\mu V/K$. More common is the typical thermopile reported in [146][143] using two different polysilicon layers (n-type and p-type differently doped) of a standard CMOS process (the gate-polysilicon and the capacitor-polysilicon) and featuring a coefficient of about 415 $\mu V/K$ [146]. Note that in this design, the parallel n-doped and p-doped polysilicon lines were connected through aluminum to prevent p-n junctions between the differently doped materials.

The thermopiles cold points were placed out of the flow stream, above the silicon substrate, not to be influenced by the temperature of the flow. The number of strips was fixed to 11 to match the microheater dimensions for an optimum sensitivity.

Three configurations were studied. The first is the more classical one as shown in Fig. 2.1 and 2.2, based on our optimized microhotplate and 2 thermopiles placed upstream and downstream of the heater at a distance of 100 μm. This configuration was realized in 2 membranes sizes, 600x600 μm^2 and 800x800 μm^2. The second (Fig. 2.3(a)) is an improvement of the first one. The microheater loop was replaced by a simple wire to reduce the power consumption and the thermopiles were placed asymmetricaly to compare the flow sensitivity in the two flow directions. The third configuration (Fig. 2.3(b)) uses the microheater of configuration 2 but the thermopiles were replaced by simple polysilicon resistors as a test. These 2 last configurations were fabricated only on 400x400 μm^2 membranes size as first prototypes.

2.2. DESIGN AND FABRICATION

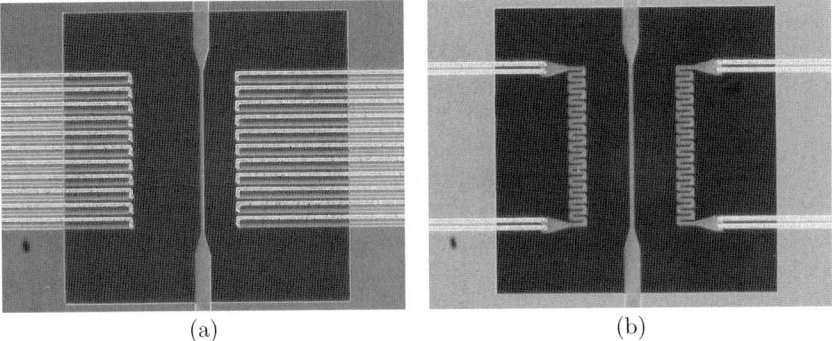

Figure 2.3: Flow sensor in the second (a) and the third (b) configuration.

The parameters of the three configurations are summarized in Table 2.1. Note that the fourth and the fifth columns show the distance between the microheater and respectively the upstream and the downstream thermopiles hot points. These distances were carefully chosen in order to allow the two thermopiles to be heated even in absence of flow. The upstream thermopile needs to be hot enough to be cooled by the flow and the downstream thermopile needs to be placed close to the microheater to sense the temperature increase of the heated flow.

Table 2.1: Parameters and critical dimensions of the three configurations.

Config.	µheater type	sensor type	Dist. th. upstream [µm]	Dist. th. downstream [µm]	membrane size [µm^2]
1	loop	thermopiles	100	100	600x600/800x800
2	wire	thermopiles	100	50	400x400
3	wire	resistors	80	80	400x400

The fabrication process exactly followed the one of the microheater, in bulk technology as well as in SOI CMOS. Fig. 2.4 shows a detailed cross-section of a flow sensor co-integrated with a transistor in SOI CMOS technology. See the previous chapter for more details. Note that the first configuration sensors were fabricated in a different batch than the sensors in the other configurations. It could explain some polysilicon doping level differences between this configuration and the other ones.

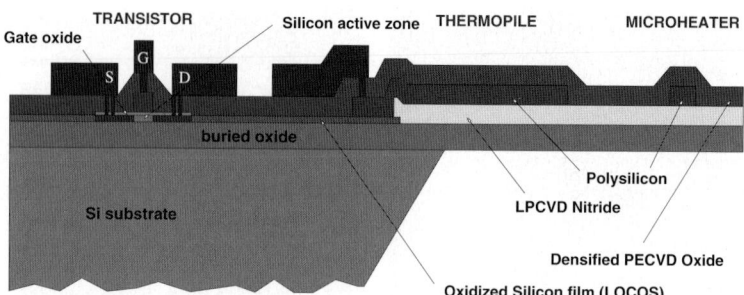

Figure 2.4: Detailled cross-section of a flow sensor co-integrated with a transistor in SOI CMOS technology.

2.3 Measurements results

2.3.1 Calibration and temperature measurements without flow

The classical microheater used in **first configuration** was calibrated in the same way as explained previously in the Chapter 1, by applying an external heating during resistance measurements. Calibrations enabled then to use the polysilicon resistors as thermometer by simple resistance measurement when heating. Results agreed very well with the model developed in the previous chapter. Increasing current was then applied on the microheater and during heating, the voltage was measured across the microheater resistor as well as across the thermopiles. No current was applied on the thermopiles to extract their voltage variation. Finally, we calculated the current-temperature relation to extract the voltage variation across the thermopiles and the microheater according to the temperature applied. These results are depicted in Fig. 2.5(a) for a 800x800 μm^2 membrane.

The voltage across the thermopiles increased when increasing the temperature on the microheater. On this process, about 140 mV were measured across the thermopiles in a 400°C ambient versus only 70 mV on a 600x600 μm^2 membrane heated at the same temperature. The thermopiles upstream and downstream of the microheater appeared well matched. A maximum offset of 6 mV appeared on the 800x800 μm^2 membrane at 400°C (1.35 mA on the microheater), and only a maximum of 2 mV on 600x600 μm^2 membranes at the same temperature (1.5 mA on the microheater). These offsets are due to unavoidable process mismatches and were taken into account in our flow rate measurements.

The microheater in the **second configuration** (as in the third one) was carefully

2.3. MEASUREMENTS RESULTS

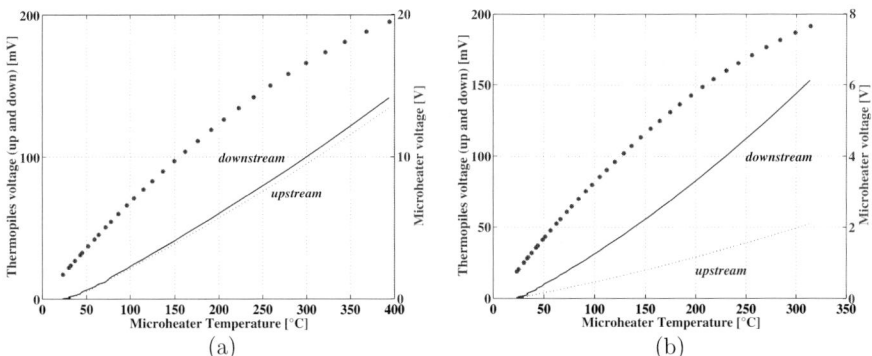

Figure 2.5: Voltage evolution across the 2 thermopiles and the polysilicon microheater versus the microheater temperature: (a) flow sensor in config. 1, membrane size 800x800 μm^2; and (b), flow sensor in config. 2, membrane size 400x400 μm^2.

calibrated in the same way to compare its power consumption to the power consumed in the first configuration. The voltage increase across the microheater and the thermopiles was also extracted when increasing temperature. Results in Fig. 2.5(b) were obtained and reveals the asymmetric positions of the two thermopiles already outlined in Fig. 2.3(a).

Finally, the power consumption was calculated in this configuration and compared with the one of the first configuration (Fig. 2.6). First of all, we observed a small power increase in both configurations in comparison with the "empty" membrane which do not support thermopiles. Thermopiles achieved higher conductivity through the membrane and lead a part of the heat to the substrate. Configuration 2 consumes less power (about 15 mW less) and offers the great advantage to be biased at only 8 V at 350°C instead of around 20 V to reach the same temperature in configuration 1.

The polysilicon resistors of the **third configuration** were finally calibrated. Fig. 2.7 shows the calibration results of 4 different sensors in third configuration. We can observe the resistance evolutions versus temperature of the 4 calibrated heater wires at the bottom of the graphics. The curves at the top show the 4 pairs of sensing resistances placed upstream and downstream of the heater.

When applying an increasing current in the microheater, the resistance variation versus temperature was extracted on the three polysilicon resistors: the microheater and the two sensing resistors. During this measurement, the two polysilicon sensing resistors were biased with 10 μA to not disturb the temperature profile. Results are depicted

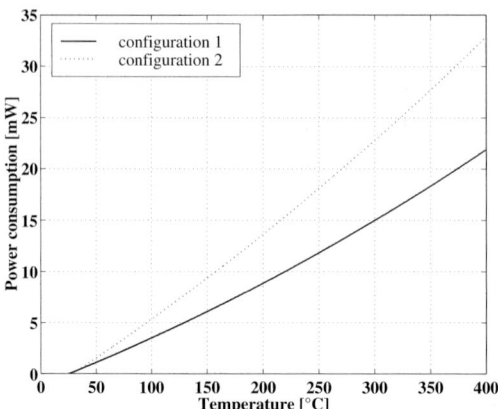

Figure 2.6: Power consumed by the microheater versus temperature. Comparison between the first (on a 800x800 μm^2 membrane) and the second configuration (on a 400x400 μm^2 membrane).

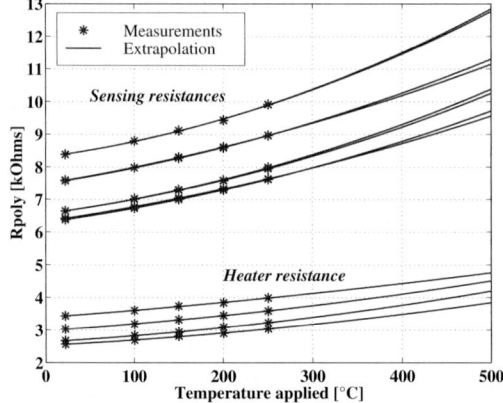

Figure 2.7: Results of calibrations achieved on 4 different sensors in the third configuration. The four curves in the bottom show the resistance increasing with the temperature in the heater wire. The 4 pairs of curves in the top give the resistance values of the resistors placed upstream and downstream of the heater.

2.3. MEASUREMENTS RESULTS

in Fig. 2.8. The third configuration has the advantage that the temperature can be accurately extracted on the flow sensing parts. Fig. 2.8 reveals that less than 180°C were dissipated through the distance of 80 μm between the hot wire and the sensing resistors. This result agrees well with our previous simulations and measurements which revealed a linear gradient of about 400°C on a distance of 200 μm.

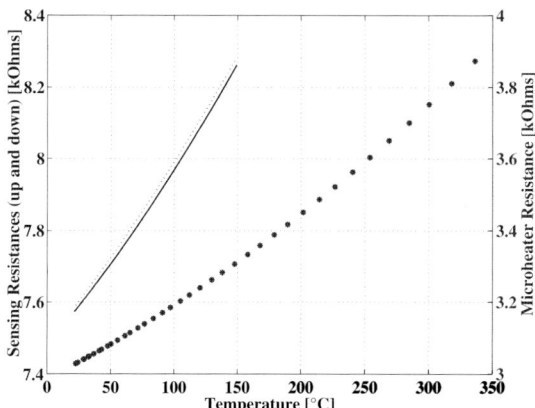

Figure 2.8: Resistance increasing of the three polysilicon resistors (the microheaters and the two sensing resistors) versus the applied temperature (config. 3, membrane size 400x400 μm²).

On the contrary, the temperature on the thermopiles (in configurations 1 and 2) cannot be extracted since we do not know the Seebeck coefficient of our doped polysilicon. By extrapolating the extracted results from our sensor in the third configuration, we estimated that our thermopiles placed at 100 μm far of the microheater reached a temperature around 180°C when the microheater was heated at about 330°C. From Fig. 2.5(b), we found that about 150 mV were measured at this temperature across the thermopiles. Knowing the number of strips N (=11) and the room temperature of the cold points (25°C), we roughly evaluated the effective Seebeck coefficient α_{Seff} of our thermopiles,

$$\alpha_{Seff} = \frac{\Delta V}{\Delta T}\frac{1}{N} = \frac{150 mV}{180°C - 25°C}\frac{1}{11} = 88 \ \mu V/°C \qquad (2.4)$$

confirming that its value is much smaller than the values obtained in [13][141] with lower polysilicon doping level. Consequently, the sensitivity S_{th} was extracted equal to 0.97 mV/°C.

172   CHAPTER 2. MICROHEATER BASED FLOW SENSOR

Reflectometry measurements could have been used to extract with more precision the temperature on the hot points of the thermopiles. An other technique was tried, consisting in inserting thermopiles up to the middle of the loop where we know exactly the temperature, to finally extract the Seebeck coefficient. Nevertheless, the narrowness of the loop accesses did not allow to put enough strips to reach an adequate sensitivity and this technique was therefore dismissed.

Finally, tests were performed to decrease the polysilicon doping level in the thermopiles in order to increase their sensitivity by increasing their Seebeck coefficient. During doping by phosphorus solid source, one wafer was removed before the saturation of impurities. Measurements on 4-points probe revealed a sheet resistance of around 160 $\Omega/\square$. Unfortunately, resistance measurements at the end of the process showed very high non-uniformities through the wafer and not reproducible results during calibration. On this wafer, positive and negative TCR were extracted through 12 polysilicon resistors in a 500 mm^2 area as shown in Fig. 2.9. Positive TCR is usually shown on conductors while negative TCR is typically the behavior of insulators or semi-conductors. The shape of the curves in Fig. 2.9 was confirmed by the literature [124].

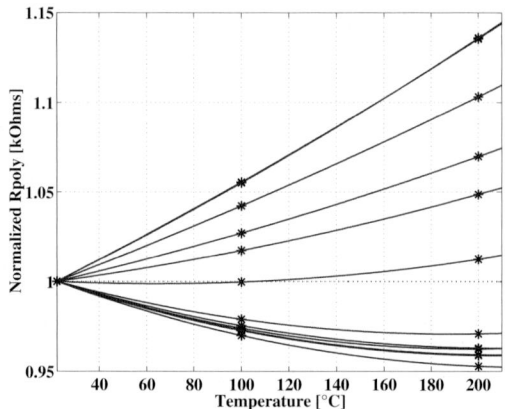

Figure 2.9: Normalized resistance change of 12 polysilicon resistors measured in a 500 mm^2 area of our wafer versus the external temperature applied on the wafer.

Even so, some measurements were tried and revealed not only an increased sensitivity for the thermopiles but also a significant power consumption increase to heat the microheater up to a target temperature. Fig. 2.9 confirms how it is hazardous to hope for an

2.3. MEASUREMENTS RESULTS

intermediate doping level from doping in solid source. The best promising method for the future would be to make the polysilicon doping by implantations. Two masks would be used in this case, one to dope the microheaters at a high level and another to dope the thermopiles with a lower dose in order to increase the sensitivity of the thermopiles without increasing the power consumption of the microheaters.

2.3.2 Microheater and thermopiles response time measurements without flow

It is critical to extract the response time of the microheater and the one of the thermopiles. In this case, the measured response time will reveal the thermal inertia of our sensor, or the time necessary for the microheater (and the thermopiles) to reach a target temperature. To measure the response time, a current pulse was applied on the microheater in order to record its behavior. A dedicated circuit was implemented to transform a typical voltage pulse into a current pulse (Fig. 2.10).

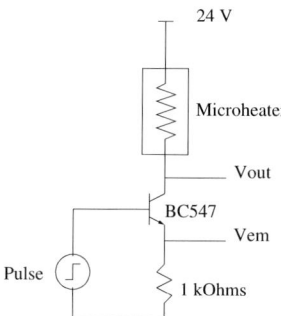

Figure 2.10: Voltage to current pulse generator used to extract the response time of our flow sensor.

The transistor current flows through the microheater and is fixed by the pulse voltage applied on the bipolar transistor base. This current was fixed around 1.5 mA (or approximately 20 V) to work around 350°C as extracted from our calibrations. For a 600x600 μm^2 membrane, V_{pulse} was therefore fixed at 2.16 V_{pp} and at 2 V_{pp} in case of a wider membrane. These first measurements were performed without flow in order to calculate the microheater and the thermopiles response time when applying heating. Results are shown in Table 2.2 and Fig. 2.11(a) and (b) for two membrane sizes.

Table 2.2: Microheater and thermopiles response time results and parameters of the measurements without flow.

Mem size $[\mu m^2]$	V_{pulse} [V]	V_{out} [V]	V_m [mV]	t_m [ms]	V_{th} [mV]	t_{th} [ms]	Power [mW]	Temp [°C]	Fig.
600x600	2.16	4.6	18.87	16	64.12	28	29	≃350	2.11(a)
800x800	2	3.8	19.15	18	103.8	30	26	≃350	2.11(b)

It can be observed in Fig. 2.11 that the voltage across the microheater increased instantaneously from 0 to 16V during the pulse and then increased slowly during the 4 last volts. In fact, a current pulse (from 0 to 1.5 mA in case of a 600x600 μm^2 membrane) is applied on a variable resistance (the microheater). During the pulse, the current increases from 0 to 1.5 mA but the resistance has not enough time to change. The resistance is equal to 9.9 kΩ at 0 mA and is the same at 1.5 mA at the end of the pulse. After the pulse, these 1.5 mA heated the polysilicon resistor and its value changed from 9.9 to 12.6 kΩ.

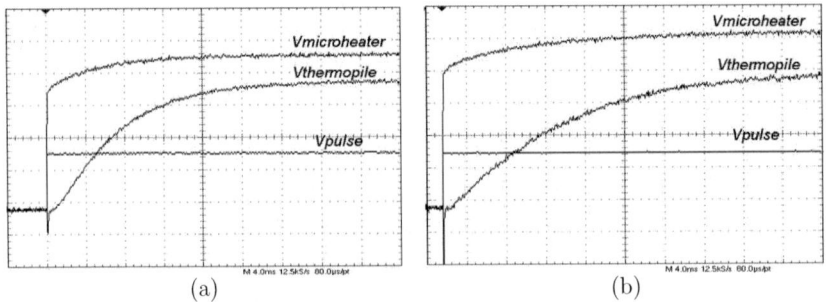

Figure 2.11: Microheater and thermopiles response time when applying heating without flow. (a) membrane size 600x600 μm^2, V_{pulse} = 2.16 V, V_{th} = 64.12 mV, V_m = 18.87 V, scale time = 4ms/square; (b) membrane size 800x800 μm^2, V_{pulse} = 2 V, V_{th} = 103.8 mV, V_m = 19.15 V, scale time = 4ms/square.

Our results show that the microheater needs only 16 ms (on a 400x400 μm^2 membrane) to raise the temperature of the microheater from room temperature up to 350°C. This value appears very good in comparison with [128] which reported a typical value of 24 ms to reach almost the same temperature. The thermopile placed 100 μm far from the microheater tooks only 12 ms more to reach its steady temperature. When the membrane area is increased, the response time is slightly increased by 2 ms.

The falling time or the time needed for the microheater to be cooled was not mea-

2.3. MEASUREMENTS RESULTS

surable by the same circuit. When the current pulse was switched off to zero, it resulted impossible to measure the resistance variation across the microheater. The falling time should be measured by switching the current in the microheater from 1.5 mA to e.g. 10 µA.

Note that the electrical RC constant of our circuit would have to be taken into account for an accurate interpretation of our results. The total parasitic capacitances can be estimated around 30 pF (including the package, the contacts of the circuit on breadboard, the coaxial wires, ...) and the microheater resistance around 10 kΩ, giving a RC constant of about 0.3 ms. This value being negligible in comparison with the calculated response time of the sensor, this appears dominated by the thermal behavior.

2.3.3 Flow measurements

Introduction

To allow flow measurements, a dedicated package was needed. Packaging was performed by inserting small gold plates under the chip in order to bring it up to the same plane as the package top surface. The sensor (and not especially the chip) was centered on the plate for an optimum symmetry under the flow.

To apply the flow selectively in one direction on the top surface of the chip, a Plexiglas channel was designed to be hermetically jointed to the package (Fig. 2.12). Measurements were performed under room temperature pure nitrogen flow[2]. As we can observe in Fig. 2.12, nitrogen was carried from two tubes into the Plexiglas channel to increase the uniformity of the flow licking the top surface of the chip.

Flow rate (in l/min) introduced in the channel was controlled upstream using a dedicated mass flow: one to measure flow rate from 2 to 22 l/min and a second one more accurate for smaller flow rates from 0 to 2.5 l/min. Flow velocity (m/s) depending on the channel section was more interesting to be known. The section of the channel being equal to 0.5 cm² (i.e.10 times wider than the tube section used to carry the flow), the relation between flow rate (F in l/min) and speed (S in m/sec) was calculated as follows:

$$S = \frac{\frac{F*10^{-3}}{60}}{0.5 * 10^{-4}} \tag{2.5}$$

---

[2]It could be very interesting to extend our measurements to hot flows (up to 300°C).

176 CHAPTER 2. MICROHEATER BASED FLOW SENSOR

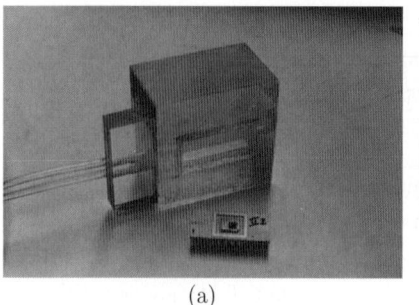
(a)

(b)

Figure 2.12: (a) Designed channel to be putted on the top of the sensor package in order to apply the flow on the top surface of the sensor. (b) Detail of the channel hermetically jointed to the chip during measurements.

Sensor in configuration 1

The measurements were performed by applying a constant current in the microheater to reach a target constant temperature. The applied power was fixed around 30 mW to reach a target temperature of about 350°C (29 mW on 600x600 μm^2 and 26mW on 800x800 μm^2 membranes). As outlined above, the thermopiles voltages were firstly measured without flow in order to extract their offset and their absolute values. These offsets were taken into account in the flow rate measurements.

Flow was applied progressively, from 0 to 22 l/min, by step of 1 l/min each 50 s in order to allow for the stabilization of the flow in the channel. The 2 to 22 l/min mass flow was used to monitor the applied flow. The measured voltage across both thermopiles is illustrated in Fig. 2.13(a) versus the flow velocity for the two membrane sizes and power consumption. The voltage difference between the two thermopiles (the upstream voltage was subtracted from the downstream voltage) is depicted in Fig. 2.13(b). This voltage difference is a flow velocity dependent signal as we can observe in Fig. 2.13(b). Furthermore, its polarity gives the flow direction.

Fig. 2.13(a) illustrates well the principle of the calorimetric sensor. When a flow passes through the sensor, it cools the upstream thermopile and its voltage decreases with the flow velocity increasing. This same flow passing above the microheater is slightly heated. Its heating strongly depends on the flow velocity. A slow flow will have enough time to be heated while a faster flow will pass above the microheater almost without being heated. This behavior can be observed in the curves of the upstream thermopile. When increasing the flow, the voltage across the upstream thermopile starts to sligthly increase

2.3. MEASUREMENTS RESULTS

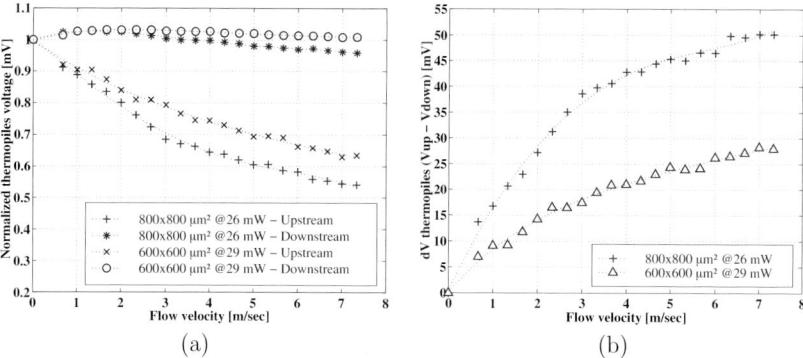

Figure 2.13: (a) Measured voltage across the both thermopiles versus the applied flow velocity for two membrane sizes (config. 1, 600x600 μm^2 and 800x800 μm^2); (b), measured differential voltage across the thermopiles versus the applied flow velocity for two membrane sizes (idem).

(i.e. heat) and then decrease (i.e. cool) when the flow velocity becomes too high to allow the flow to be heated by the microheater. Comparing now the two membrane sizes and the different injected power, we see that the described behavior appears more or less significant depending on the temperature of the microheater and the membrane size.

flow velocity from 1.5 to 22 l/min (i.e. from 0.5 to 8 m/sec or from 1.8 to 28.8 km/h) were measurable with a great accuracy. The more accurate mass flow (0 to 2.5 l/min) was used to explore the sensitivity to lower flows but did not provide reproducible results for flow rates lower than 0.5 m/s (or 1.5 l/min) probably due to some turbulences caused by our imperfect measurement channel. Higher flow velocities could be measured since at 8 m/s, the curve does not appear saturated yet. Unfortunately, we did not dispose of a higher range mass flow.

Measurements were also performed when changing the power applied on the microheater to compare the resulting sensitivity (Fig. 2.14).

When increasing the flow, we observed a slow decrease of the microheater resistance due to its progressive cooling. The normalized microheater resistance variation was extracted during our measurements and is depicted in Fig. 2.15. Small variations of only 3 % are observed on the whole flow velocity range. When switching off the flow, the microheater resistance came back to its original value without flow confirming the perfect reversibility of our sensor.

These results observed on the microheater revealed that a constant heating current

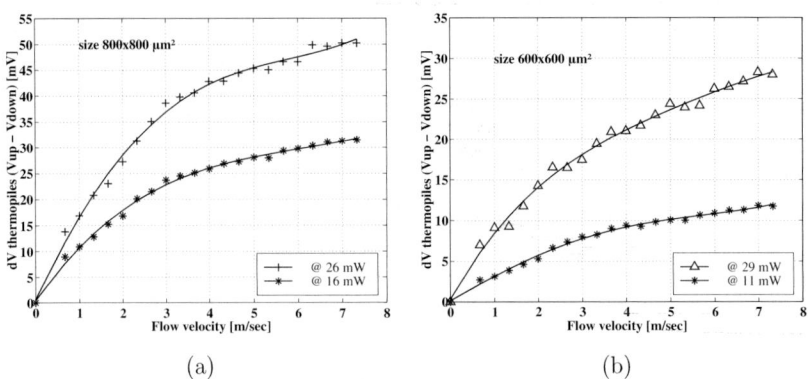

Figure 2.14: Differential voltage across the thermopiles versus flow velocity when changing the power and therefore the temperature applied on the microheater: (a) for a 600x600 μm^2 membrane size and (b), 800x800 μm^2 membrane size (config. 1).

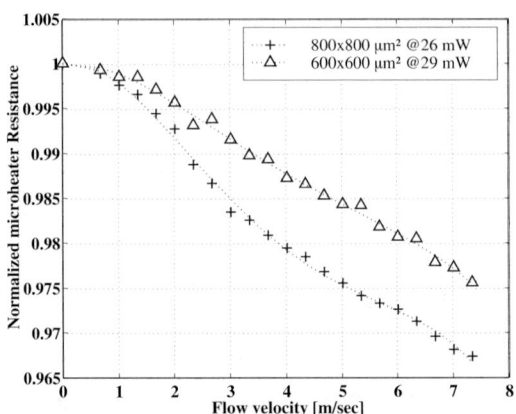

Figure 2.15: Normalized microheater resistance decreasing with the flow velocity for both membrane sizes and power consumptions (flow sensor in configuration 1).

2.3. MEASUREMENTS RESULTS

(or voltage) does not automatically provide a constant heating power. This observation was also reported in [143] and they suggested to connect another resistor (featuring the same value) in series to the heater, on the bulk silicon, in order to maintain constant the power consumption. This additional resistor does not change its resistance because it is not thermally isolated on a membrane. This compensation revealed promising results as reported in [143] but increases the power consumption. Furthermore, it can compensate the polysilicon microheater resistance relaxation at high temperature versus time as reported in the previous chapter.

Table 2.3: Thermopiles response time results and parameters of the measurements in presence of flow.

Mem size $[\mu m^2]$	V_{DC} $[V]$	V_{out} $[V]$	V_{th} $[mV]$	t_{th} $[ms]$	Power $[mW]$	Temp $[°C]$	Fig.
600x600	2.16	4.6	64.12	**25**	29	330	2.16(a)
800x800	2	3.8	103.8	**40**	26	330	2.16(b)

Finally, response time measurements were performed under flow. In this case, the response time will reveal the time needed by the sensor to give a stable response to a rough flow variation. For this measurement, the V_{pulse} used in absence of flow was replaced by a continuous voltage (V_{DC}) as we can observe in Table 2.3. Flow was pulsed from 0 to 22 l/min (voltage decreasing on the thermopiles) and from 22 to 0 l/min (voltage increasing on the thermopiles). The response time was observed the same going up and going down. Thermopiles appeared to take 25 ms to detect a sudden debit variation of 22 l/min on a 600x600 μm^2 membrane but 40 ms on a 800x800 μm^2 one. A trade off appears between sensitivity and response velocity.

Sensor in configuration 2

Sensors in configuration 2 were measured under increasing flow velocity in both directions (see Fig. 2.3(a)) from left to right when downstream thermopiles were closer to the hot wire than the upstream ones and from right to left (when upstream thermopiles were closer to the microheater) to compare the sensitivity to flow velocity in both cases. Current was fixed equal to 2 mA across the heating resistor to reach a temperature of ∼300 °C and consuming a power of 15 mW. Both results are compared in Fig. 2.17. As depicted in Fig. 2.17(a), it is observed that the sensitivity of the downstream thermopiles was increased by a few percents when getting them closer to the microheater. Since the microheater area is

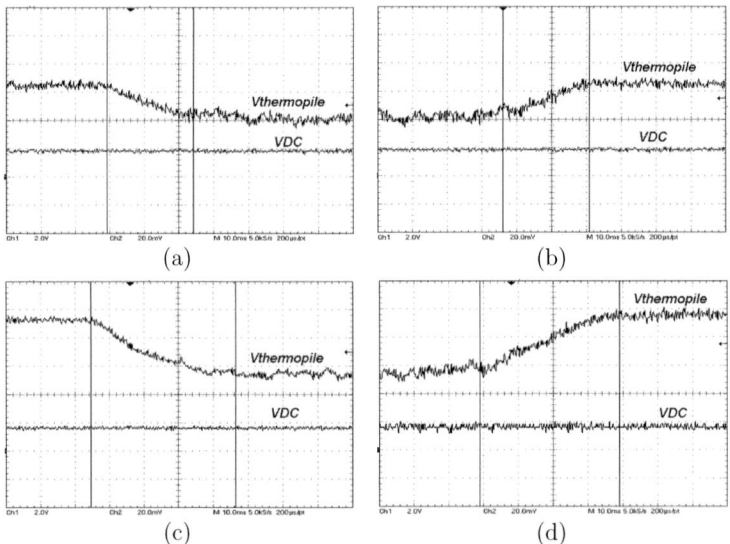

Figure 2.16: Microheater and thermopiles response time when switching on the flow (to the left) and when switching off the flow (to the right): (a) and (b) Membrane size 600x600 μm^2, V_{DC} = 2.16 V, V_{th} = 103.8 mV, scale time = 10 ms/square; (c) and (d) Membrane size 800x800 μm^2, V_{DC} = 2 V, V_{th} = 64.12 mV, scale time = 10 ms/square.

2.3. MEASUREMENTS RESULTS

very small and the downstream thermopiles are closer to the microheater, the flow does not have enough time to be heated and therefore, to cool the downstream thermopiles. On the contrary, when the downstream thermopiles are further from the microheater (with reverse flow, from right to left), the flow has more time to be heated when passing above the membrane and slightly increase their temperature. Regarding the upstream thermopiles, it appears that left to right measurements reveals a higher cooling of them than right to left measurements due to their higher distances to the microheater in the first case than in the second one.

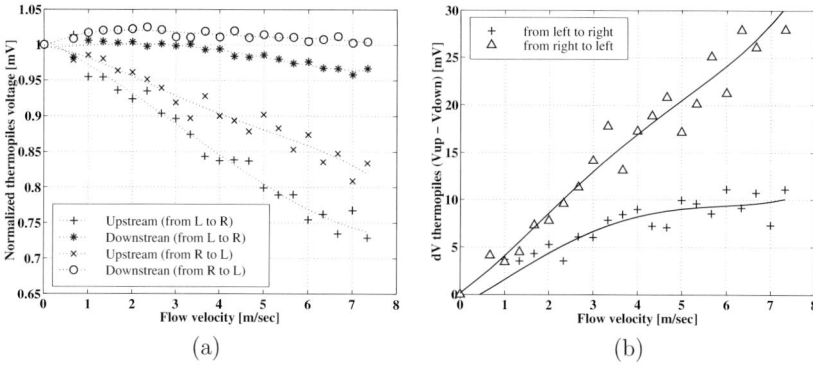

Figure 2.17: (a) Measured voltage across both thermopiles versus the applied flow velocity with a flow passing from left to right and from right to left (config. 2, membrane size 400x400 μm^2); (b), measured differential voltage across the thermopiles versus the applied flow velocity with a flow passing from left to right and from right to left (idem).

Fig. 2.17(b) depicts the voltage difference between the two thermopiles (the upstream voltage was subtracted from the downstream voltage) in both flow directions. First of all, we observe that the asymmetric arrangements of the thermopiles change the shape of the output characteristics as reported in [140]. This graph reveals that, regarding the voltage difference between both thermopiles, a higher sensitivity to flow velocity is obtained for flow passing from the right to the left. Despite of the lower sensitivity measured in each separate thermopile when applying the flow in this direction, the difference reveals a higher sensitivity thanks to the higher voltage value on the more sensitive thermopile (the upstream one). Therefore, it is clearly better to put the upstream thermopiles closer to the microheater than the downstream ones to shift the saturation to higher flow velocities.

Ageing tests were done on flow sensors in configuration 2 to study the effect of the microheater polysilicon resistance relaxation on the voltage measured across the thermopiles. Firstly, we achieved measurements without flow during the time which was required to make the previous measurements in presence of flow. A fresh sensor never heated was used for this test. At the end of the 18 minutes of heating at a constant current of 2 mA, we measured voltage drifts of about 5 % on both thermopiles and of about 3.5 % on the microheater (Fig. 2.18(a)). Increasing voltages were observed on the microheater as well as on both thermopiles resulting in temperature increasing which confirms the observations discussed in the previous chapter when constant current is applied in the microheater.

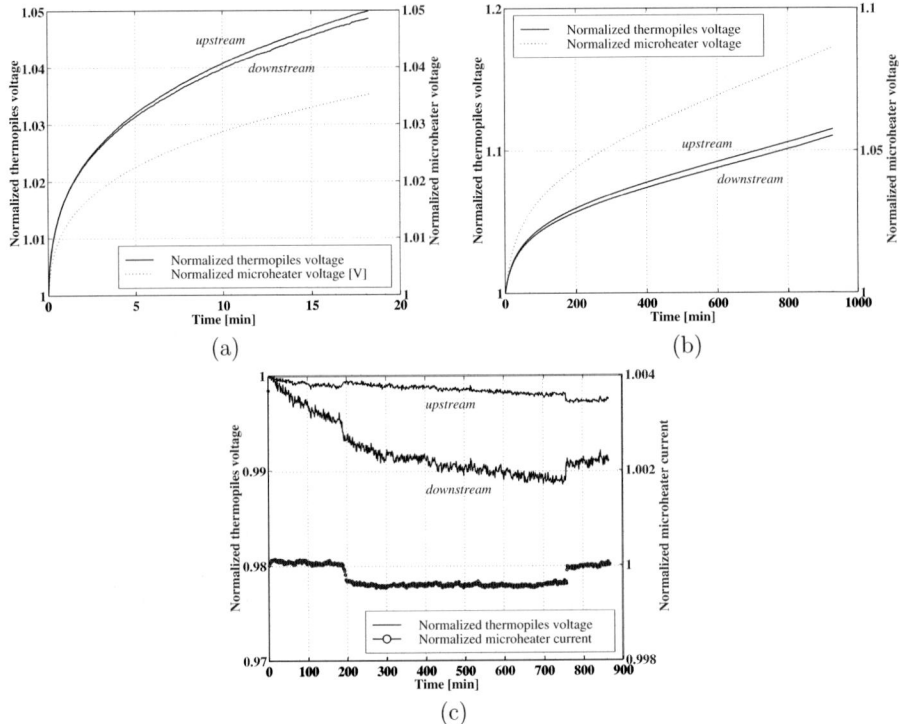

Figure 2.18: Normalized voltage relaxation versus time through the heating microheater and the two thermopiles; (a) heating of a fresh sensor at a constant current of 2 mA during the first 18 min; (b), heating of an other fresh one under the same conditions during 900 min; and (c), heating of an other fresh sensor at a constant voltage (supplying the same power) during 900 min.

Similar measurements were performed for a longer period of time (900 min) using a fresh microheater. The results (Fig. 2.18(b)) show that the voltage begins to strongly increase and tends to be slowly stabilized with time. On the contrary, a fresh microheater

2.3. MEASUREMENTS RESULTS

pre-heated for about 2 hours shows a voltage drift across the thermopiles of only 5 % after 13 hours, which can be negligible when high sensitivity is reached. So for a better accuracy and reproducibility, it is therefore required to perform a membrane pre-heating at its operating temperature during 2 hours at least before the sensor use.

An other technique would be to perform flow measurements at constant voltage. We concluded in the previous chapter that this allows to sligthly decrease the temperature drift during measurements at the operating temperature. Ageing tests were done at a constant voltage calculated to supply the same power than a constant current of 2 mA. Results are depicted in Fig. 2.18(c) and confirm the very slight temperature decrease over time. The power consumption was almost constant during the measurements as confirmed by the constant current value. Note that the two vertical drifts in our curves were caused by the measurements facilities (the vacuum supporting the wafer was switched off on the first jump and switched on on the second one). The voltage across the thermopiles decreased by only 1 % after 900 minutes in comparison with the 11 % measured at constant current. Nevertheless, the voltage difference between the upstream and the downstream thermopiles remains almost the same in both cases.

During measurements at constant current, a small voltage increase across the thermopiles was observed at very low flows (< 2 m/s) when the flow was not high enough to compensate this increase. It is the case in Fig. 2.17(a) where a small bump appears below 2 m/s. At higher flows, it becomes fully negligible. To confirm that the polysilicon resistor relaxation can be neglected, measurements were done when increasing the flow from 0 to 8 m/s and compared with measurements when decreasing the flow. Both measurements gave exactly the same results with great accuracy. It confirmed that the microheater relaxation has no impact on the flow measurements in configuration 2 thanks to the high sensitivity achieved.

Finally, ageing tests were performed during 2 hours in presence of a 8 m/s flow and the voltage then appeared constant across both thermopiles as well as across the microheater. We thereby confirm that the air flow passing above the polysilicon wires contributes to offset the small resistance relaxation.

In configuration 2 (as well as in the third one), direct measurements on the microheater revealed less interesting results than in the first configuration. A resistance variation of only 1 % was measured on the flow velocity range (Fig. 2.19). Furthermore, the polysilicon resistance relaxation firstly leads to a small resistance increase between 0 and 1 m/s while the flow was too small. Later, the flow was high enough to compensate the relaxation and

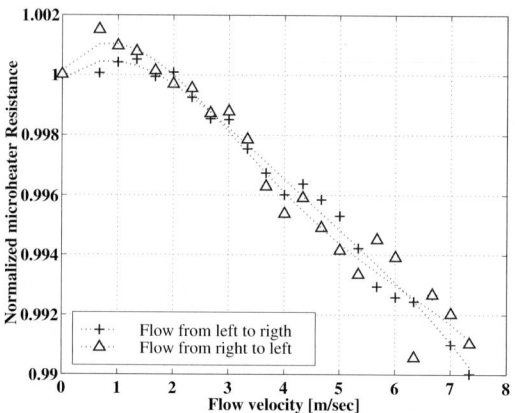

Figure 2.19: Normalized microheater resistance decreasing with the flow velocity for 2 directions of flow. Flow sensor in configuration 2.

the resistance started to slowly decrease. In comparison with configuration 1, the heated area is too small in these two configurations and does not provide an area large enough to be cooled by the flow.

Sensor in configuration 3

In configuration 3, the sensitivity was measured to be very small due to the too low temperature of the sensing resistors. 2 mA (or ~300°C) was applied on the microheater and without flow, a temperature of around 150°C was measured on the sensing resistors by applying a current of 10 μA.

In presence of flow, the resistance variation of the downstream resistor was observed negligible while the upstream resistor showed resistance variation of only 2 % on the whole measured flow velocity range (Fig. 2.20(a)). This upstream resistance decrease leads to a weak temperature decreasing of about 5°C. In this configuration, the flow velocity was expressed as the voltage difference between the upstream and the downstream resistors as shown in Fig. 2.20(b).

2.3. MEASUREMENTS RESULTS

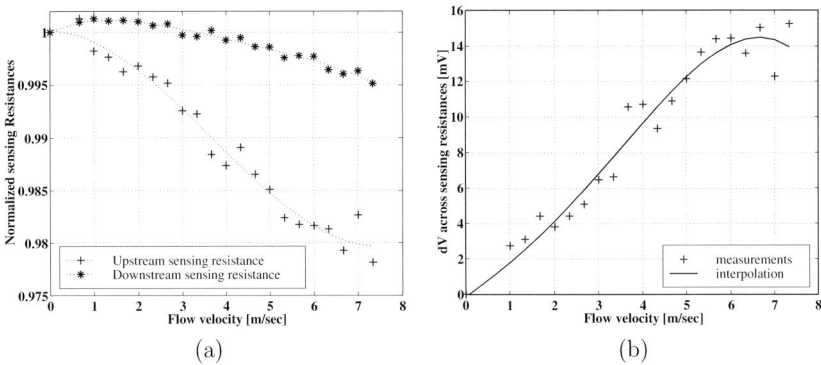

Figure 2.20: (a) Measured voltage across the both resistors versus the applied flow velocity; (b), measured differential voltage across the resistors versus the applied flow velocity (membrane size 400x400 μm^2).

Discussions and summary

A comparison of each configuration was done and is depicted in Fig. 2.21. The best option for high sensitivity on a large flow velocity range and low power consumption is the first configuration with the largest membrane (800x800 μm^2). Nevertheless, this configuration suffers from lower robustness due to its high membrane size and from high voltage bias (around 15 V to reach 16 mW). This first configuration also shows a higher sensitivity to low flows but slowly begins to saturate at velocity of about 7 m/s. Regarding the other configurations, the second configuration with its upstream thermopiles closer to the microheater than the downstream ones, shows interesting characteristics. Only 7.6 V is necessary to reach 15 mW, and its membrane size is only 400x400 μm^2 which provides a higher robustness. Furthermore, the characteristic does not saturate up to 8 m/s and reveals the same voltage difference value than the first configuration, i.e. about 30 mV at flow velocity around 7.5 m/s. Nevertheless, its sensitivity to low velocities seems not so good and due to the lower sensitivity of each separate thermopile in this case, the results are noisier.

Regarding the sensitivity of the microheater itself to flows, it clearly appears that the microhotplate of the first configuration offers a higher area in contact with the flow to be cooled. On the contrary, the small wire of the second and third configurations is not large enough.

About the response time results in presence as well as in absence of flow, the following

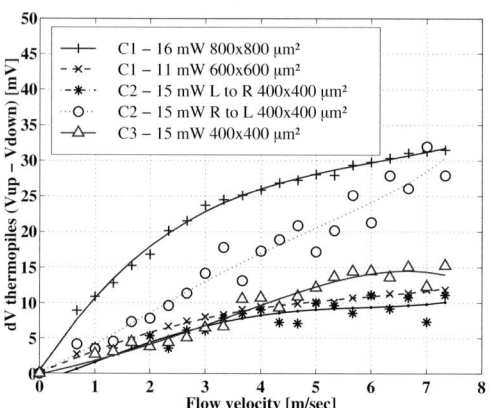

Figure 2.21: Measured differential voltage across the thermopiles (or the resistance in config. 3) versus the applied flow velocity in each configuration (C1 to C3) and some variable parameters (membrane size for C1 and flow direction for C2).

conclusions can be drawn: the microheater on a 600x600 μm^2 membrane takes 16 ms to reach 350°C, and a thermopile placed 100 μm far from the microheater takes 12 ms more. On a larger membrane (800x800 μm^2), the time is a little bit increased (by 2 ms) but for a better sensitivity under flow. When the flow is switched on, thermopiles takes more time on 800x800 μm^2 membrane (15 ms more than on 600x600 μm^2) to detect a sudden flow rate variation of 22 l/min due to the larger voltage decrease. But finally, larger membranes yield a better sensitivity. These results reveal that there is a trade off between sensitivity and high response speed.

So in summary, the first configuration (with membrane size of 800x800 μm^2) remains the best option to reach a good sensitivity to flows in the range from 0 to 8 m/s. Beyond this value, the characteristic begins clearly to saturate. On the contrary, the second configuration (with the closest thermopiles on upstream) offers the best option for higher flow velocities which can be evaluated with lower accuracy. It is not so well adapted to low flow velocities but its small size (400x400 μm^2) allows its use to measure harsh flows. Finally, the third configuration was very interesting as demonstrator and to illustrate new thermal properties of our dielectric membrane.

Although, some improvements could be done in order to further increase the sensitivity as well as the measurements range for a given power consumption by placing the thermopiles closer to the microheater, especially in the first configuration.

2.4 Discussions and comparison with the state-of-the-art

A review of the characteristics of recent flow sensors realizations is summarized in Table 2.4. Some explanations are needed for a good understanding and a correct interpretation of the full set of datas.

Firstly, it must be noted that the calorimeter principle chosen for our design is not an exclusive strategy to measure the flow with a thermal device. As reviewed in [147], the anemometer and the thermal time-of-flight (so called TOF) are also common in such sensors:

- **Anemometer (heat loss)**: when a heated membrane is exposed to a moving fluid, the wall heat transfer rises with rising flow. The velocity is calculated from the electrical power needed to heat the membrane to a given temperature (constant temperature anemometer) or from the temperature reached with a given power (constant power anemometer). By this method, the velocity at the point of the membrane can be measured [147].

- **Thermal time-of-flight (TOF)**: a short impulse is applied on a microheater. The thermometer measures the time needed to the pulse to cover a given distance [147].

An anemometer as well as a calorimeter is characterized by its constant power applied in the microheater. On the other hand, the TOF-based flow sensor is characterized by its impulse power (50 mW in [147] for instance). Note that the interesting design reported in [147] features a combination of two detecting methods (calorimeter and TOF) which results in a considerable increase in measuring range.

We can observe in Table 2.4 that independently on the sensor principle, the structure of the sensor can be based either on a membrane, a bridge or an other structure (such as porous silicon for instance in [148] or the silicon substrate in [149]). On the other hand, the transducer of the sensors presented here are either thermopiles or resistors. We integrated polysilicon resistors to guarantee CMOS compatibility but Platinum (or Germanium in [140]) is more often used thanks to their higher temperature coefficient of resistance (TCR). Note that diodes are also reported as transducer in [150] and [151] for liquid flow measurements.

The CMOS compatibility of a sensor design was judged in term of its ability to be fabricated in a standard CMOS process without special care and complicated steps. Based on these criteria, porous-silicon-based, Ge-based, Pt-based or gold-based sensors were rejected from our "CMOS compatible" appellation. Note that other fully CMOS compatible designs, not reported in this Table, can be found in literature and reveal very interesting properties. We can especially note the works of [145] using highly sensitive n-poly/p$^+$-silicon thermopiles, and those of [146][143] featuring flow velocity range from 0 to 38 m/s with a total power consumption of only 3 mW with thermopiles based on two different polysilicon layers. Finally, we can observe in [149] a typically CMOS compatible process since the flow sensor is directly integrated on the silicon substrate without special structure. Nevertheless, the non isolated microheater revealed very long response time (of about 5 sec). Finally, the design of the anemometer reported in [152] is not CMOS compatible since it uses Cr-gold layers as bonding pads.

We could precise some additional comments about Table 2.4. The reported size expresses the size of the structure only (not of the whole sensor), i.e. its width times its length, its thickness appearing at the line below. The heater temperature ΔT_{heat} expresses the temperature difference between the heater and the room temperature. Two values of sensitivities were reported. The first one expresses the sensitivity measured on the characteristic at very low flow (close to zero) while the second one reports the sensitivity at the end of the flow range. The "mesh-membrane" reported in [153] shows etching openings in its structure in order to allow the undercutting of the silicon substrate. This design reduces the etching time and still provides stiff-enough suspension.

2.5 Conclusions

First of all, our measurements without flow improved our thermal study about the microheater as well as about the membrane for a better understanding of their behavior. Our thermopiles in configurations 1 and 2 and the polysilicon resistors in configuration 3 allowed to measure the thermal gradients through the membrane and to confirm our simulations results. The response time of the microheater and the sensing thermopiles was also extracted since it is especially critical in flow sensing applications.

Our measurements in presence of flow revealed interesting behaviors, from one configuration to an other. Measurements such as voltage variations versus flow velocity, ageing tests, reliability tests at high flows, response time to abrupt flow variations, comparison

2.5. CONCLUSIONS

between results when increasing and decreasing the flow velocity, and resistance variations of the microheater under flow were performed. The comparison between the three configurations revealed that the second configuration showed a higher sensitivity to a higher flow velocity range when the upstream thermopiles were placed close to the microheater.

Comparing our measurements results with the recent published results the following observations can be done. First of all, a really high temperature was reached on the heating resistor for a very low power consumption which could provide a great sensitivity to flows. Unfortunately, our p^{++}-polysilicon/aluminum thermopiles did not provide a fair sensitivity, but gave the advantage to make our design simple and fully CMOS compatible. A better sensitivity could be provided by doping differently the thermopiles and the microheater using a second implantation mask, as confirmed by our preliminary tests. In this case, the microheater would be doped at the same time as the transistors gate. So, regarding the CMOS compatibility, our sensor offers good performances for a very simple design thanks to the quality of our 1 μm thin dielectric membrane. Furthermore, our sensor offers a good sensitivity over a large flow velocity range, from the low flow velocity to the higher one, at a low power consumption, and a response time in the average of the state-of-the-art. Therefore, our flow sensor seems to present attractive trade-offs between the above-mentioned parameters for a large range of applications.

Figure 2.22: Four directions flow sensor with its 4 thermopiles and its microhotplate with 4 accesses for symmetry.

We reported flow measurements in only one direction. The voltage difference between the downstream and the upstream thermopiles appeared to be a flow velocity dependent signal. In the same way, its polarity gave the flow direction, from left to right or from right to left. As suggested above, it could be very interesting to extend our design in order to measure in two directions perpendicular to each other, or even over the full 0° - 360° angle. The design and the fabrication of such sensors were done as we can observe in Fig. 2.22 but were not yet characterized. The measurements under flow of this kind of sensors is really complicated to implement. Furthermore, calculations need to be done in order to extract the flow direction from the results extracted on the four thermopiles. An example of such sensor can be found in [154].

Finally, a discussion could be done about the ability of our device to sense hot flows. If the temperature of the flow increases, the sensitivity of our sensor will decrease since the temperature difference between the flow and the microheater will decrease. To make our sensor insensitive to any flow temperature change, the following improvement could be proposed [155]. Four polysilicon resistors could be mounted in Wheastone bridge. One of them (R_1) would be integrated close to the microheater and the three others (R_2 to R_4), featuring the same value as R_1 heated at the operating temperature, on the substrate, at the room temperature. Since the variation of R_1 would be slower than the variation of the other resistances under hot flow, the measured voltage across the bridge would follow the temperature variation between the flow (R_2 to R_4) and the microheater (R_1). This measured voltage amplified and applied on the gate of a transistor would lead to an increasing current supplying the microheater (if it is connected to the drain of the transistor), proportionally to the flow temperature increase.

2.5. CONCLUSIONS

Table 2.4: Review of some recent published flow sensors results. See additional comments in the full text.

Sensing principle	Transducer	Structure and size	Flow rate range	Sensitivity	Year and Ref. IC compatibility
Calorimeter P=cste=13mW -	Thermopiles poly/poly 4.05 mV/K	Membrane 400x400 μm^2 x 3 μm	0 - 900 sccm	0.36 mVmW^{-1}sccm^{-1} 0.034 mVmW^{-1}sccm^{-1}	1993 [143] Yes
Anemometer P=3.3V x 1.8mA ΔT_{heater} = 80 K	Polysilicon -	Bridge - x 1.25 μm	0.005 - 35 m/s	22 mV(m/s)$^{-1/2}$ @T=cste 5 mV(m/s)$^{-1/2}$ @I=cst	1996 [152] No
Calorimeter P=cste=1mW ΔT_{heater} = 5 K	Thermopiles Bi$_{0.87}$Sb$_{0.13}$/Sb 13.5 mV/K	Membrane 3.6x1.15 mm^2 x 0.8 μm	0 - 2.2 m/s	5.9 VW^{-1}m^{-1}s 1 VW^{-1}m^{-1}s	1998 [156][157] No
Calorimeter P=cste=67mW -	Thermopiles al/poly -	Porous silicon 240x1170 μm^2 x 40 μm	0 - 0.4 m/s	6 mVW^{-1}m^{-1}s 6 mVW^{-1}m^{-1}s	1999 [148] No
Calorimeter P=cste=5mW TOF (50mW)	Thermopiles poly/gold 6 mV/K	Membrane 600x600 μm^2 x 0.15 μm	0.125 - 8 mm/s 12 - 140 mm/s	- -	1999 [147] No
Anemometer V=cste=4V I=cst=12.96A	Pt resistors	Mesh-membrane 600x600 μm^2 x 0.2 μm	1.5 - 11 m/s	0.046 mA(m/s)$^{-1/2}$ 7.98 mV(m/s)$^{-1/2}$	2000 [153] No
Anemometer P=cste=500mW ΔT_{heater} = 24 K	Thermopiles p$^+$-Si/al 6 mV/K	Si substrate 4x3 mm^2	2 - 18 m/s	- -	2002 [149] Yes
Calorimeter P=cste=2mW ΔT_{heater} = 460 K	Pt resistors	Bridge - x 1μm	0 - 2 m/s	2 VW^{-1}m^{-1}s 0.4 VW^{-1}m^{-1}s	2002 [158] No
Calorimeter P=cste=4mW ΔT_{heater} = 23 K	Ge resistors	Membrane 0.5x1.1 mm^2 x 0.8 μm	0.01 - 200 m/s	- -	2003 [140] No
Calorimeter P=cste=15mW ΔT_{heater} = 300 K	Thermopiles al/poly 0.97 mV/K	Membrane 400x400 μm^2 x 1 μm	0 - >8 m/s	0.27 VW^{-1}m^{-1}s 0.27 VW^{-1}m^{-1}s	2003 Thesis config. 2 Yes

Chapter 3

Gas Sensors on microhotplate

3.1 Introduction

As outlined in the first chapter of this part, our developed microhotplate is especially dedicated for gas sensing purposes. After decades of neglecting the effects that ever-increasing industrial, agricultural and military activities can have on human health and ecosystems [20], there is an urgent need nowadays to dispose of sensing devices having the assignement to improve the environmental and safety control of toxic gases. There is also a great need for this kind of sensors for optimizing combustion reactions in the emerging transport industry, as well as domestic and industrial applications [20].

The semiconductor gas sensors constitute one of the most interesting gas sensor devices. They are small, portable, consume little power and are cheap which allow their use in domestic applications as well as in reduced spaces (as car engine for instance) [20]. The principle of this kind of gas sensors was thoroughly introduced in Chapter 1 of this part. We explained that a microhotplate featuring low consumption, high thermal uniformity and high robustness is the basic cell of a semiconductor gas sensor to allow the reaction with gases to take place through the heating of a sensitive metal oxide layer. Furthermore, we emphasized the fact that its fabrication needs to be fully compatible with CMOS-SOI technology in order to increase its fabrication easiness and to decrease its cost. Finally, we saw that our microhotplate, when previously calibrated, allows to sense its operating temperature in same time that it is heated.

To obtain a gas sensor, a sensing material which reacts with the gas atmosphere by adsorption reactions on its surface, must be added on the uniformly heated area. This layer is commonly a metal-oxide layer such as SnO_2 or WO_3. As its conductivity changes

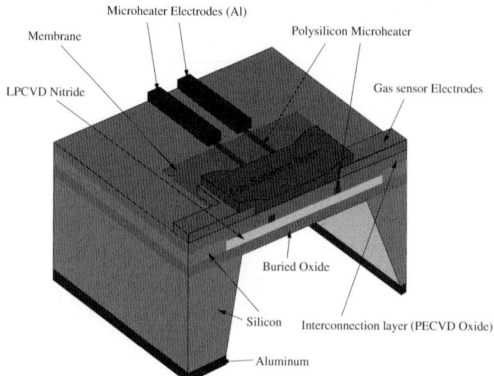

Figure 3.1: Fully integrated smart gas sensor concept in SOI-CMOS technology.

in presence of gases, metallic interdigitated electrodes must finally be added underneath the sensing layer in order to sense its resistivity modulation. A schematic of a complete gas sensor is depicted in Fig. 3.1.

Therefore, several post-processing steps are further needed on the finished microhotplate in order to obtain a co-integrated gas sensor. First of all, interdigitated electrodes need to be deposited onto the densified PECVD oxide of the membrane. It will be detailed in the next lines that their design as well as the choice of their material is critical to reach a high sensitivity and to allow CMOS co-integration and high temperature operation. Next, the metal oxide sensitive layer must be deposited. Three methods of deposition were investigated: sputtering and two other methods allowing to deposit thicker layers, drop coating and screen-printing techniques. Depending on the method of deposition, this step is performed either after the backside TMAH etching and before the dicing and packaging (for sputtering), or after the sensor packaging (for drop coating), or before the TMAH etching and packaging. All of these steps will be described further. After the post-processing description, measurement results will be presented, firstly without gas and followed by measurements under different gas ambients.

3.2 Interdigitated electrodes: from design to deposition

3.2.1 Design

The gap width between interdigitated electrodes is critical to achieve a good sensitivity. The height of the electrical field lines will depend on this gap width: a thicker sensitive layer will require higher electrical field lines to sense up to the layer surface and therefore, a wider gap. The gap width can also have an impact on the selectivity between gases. Furthermore, since the metal oxide resistivity decreases when the temperature increases, a higher sensitivity can be obtained with higher measured resistances. Therefore, if the gap width between fingers is increased while the number of fingers is decreased, the measured resistance will be higher as well as the sensitivity at higher temperatures.

All studied designs are reported in Table 3.1. First design (entitled "old" in Table 3.1) used small fingers gap and a high number of fingers. With gold in a thickness of around 100 nm as constitutive materials, the measured resistance at 250°C was too small (around 200 Ω for a 10 μm thick drop coated layer) to reach a good sensitivity at high temperatures. Therefore, this design was replaced by 3 optimized ones as reported in Table 3.1, each of them being based on a different fingers gap. These designs revealed better results as it will be shown further.

Table 3.1: Designed parameters of the interdigitated electrodes.

Design	fingers width [μm]	fingers gap [μm]	number of fingers pairs
old	3	3	17
new 1	10	54	2
new 2	10	18	4
new 3	7	6	8

In the future, it would be interesting to optimize the fingers design according to the chosen metal for the interdigitated electrodes and to the type of metal oxide used as sensitive layer in order to reach a typical resistance of about 100 kΩ. It seems a good compromise between high sensitivity and low consumption.

3.2.2 Metal selection and deposition

It was explained above that the type of metal chosen for our interdigitated electrodes as well as its thickness have an impact on the measured resistance. This metal needs also to follow some requirements which will dictate our choice:

Process aspects:

- able to be patterned by lift-off (in thicknesses up to 400 nm);

- featuring high adhesion to densified PECVD oxide, directely or via an adhesion layer (such as Cr or Ti). The adhesion layer needs to fill the same conditions;

- inert in pure TMAH or in TMAH with additives (Si powder and APS powder);

- noble non oxidizable metal to avoid contact problems with metal oxide layer;

Thermal aspects:

- featuring high melting point to allow heating as high as 700°C without damage;

- featuring intermediate thermal conductivity (neither to high to avoid to heat the substrate, nor to low to decrease the thermal uniformity);

Electrical aspects:

- able to be wire bonded for packaging (or giving good contact with aluminum to allow its use as a transition layer);

- CMOS compatible;

- featuring an indifferent resistivity at ambient temperature and TCR since the resistance of the interdigitated electrodes will be negligible in comparison with the metal oxide resistivity.

To make our choice easier, Table 3.2 summarizes the main physical, thermophysical and electrical properties of most metals commonly used in microelectronics [159][160].

All of our first fabrications were performed with **Gold** and chromium as adhesion layer. This metal is noble, cannot be oxidized and is known to be inert in a lot of chemicals, especially in TMAH. 7 nm of chromium followed by 100 nm of gold were evaporated on our densified PECVD oxide layer and were easily patterned by lift-off (in fuming nitric

3.2. INTERDIGITATED ELECTRODES: FROM DESIGN TO DEPOSITION 197

Table 3.2: Summary of the main physical, thermophysical and electrical properties of most metals used in microelectronics.

Metal	Melting point [$°C$]	Thermal exp. coeff [$10^{-6}/°C$]	Thermal conductivity [$W/m\ K$]	Electrical resistivity [$\mu Ohm*cm$]	TCR [$10^{-3}\ °C^{-1}$]
Aluminum	660	23	240	4.3	4.2
Tungsten	3415	4.5	200	5.5	4.8
Chromium	1900	8.2	66	20	5.9
Titanium	1668	8.5	22	47.8	3.8
Gold	1063	14.2	297	2.2	4
Platinum	1774	9	71	10.6	3.8

acid). Chromium and gold proved very robust in TMAH and showed high adhesion to the oxide layer.

Nevertheless, gold revealed a lot of drawbacks. First of all, direct wire bonding on gold was not so convincing and leads to high resistive contacts. Therefore, our idea was to provide large aluminum pads on the future location of the gold wires. By this way, gold-aluminum interconnections were obtained after gold evaporation and wire bonding was performed on aluminum instead of gold. We measured good contacts at room temperature but when increasing temperature on the whole wafers, strange colors were observed and some migrations were observed between both layers. [160] confirmed that Au_xAl_y alloys are extremely brittle and that some reactions between both layers can destroy the electrical contacts over time. Nevertheless, electrically, the contacts seemed to be undamaged in our case despite of these observations.

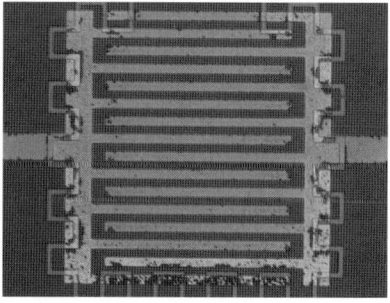

Figure 3.2: Cracks in interdigitated gold electrodes when heated at around 700°C during 1h00 (600x600 μm^2 membrane).

Another main drawback of gold is its too low melting point. When heating the microheater at 700°C, gold melted slowly as shown in Fig. 3.2. Picture was taken after heating during 1 hour around 700°C (2 mA) on a 600x600 μm^2. A detailed observation showed irreversible cracks on gold layer (but the chromium adhesion layer seemed to be intact). Nevertheless, this problem was reduced when electrodes were covered with the gas sensitive layer. [128] confirms our observations by the following explanation: above 650°C, the Cr/Au contacts undergo to a progressive degradation which is due to the chromium grain-boundary diffusion through the gold layer towards the surface.

Finally, as it will be detailed in the last chapter of this part, the e-beam metal evaporation induces X-ray damage on electronics devices. These damages are revealed as threshold voltage drifts (to the lower values in case of nMOSFETs) and increasing leakage currents due to positive trapped charges in the gate oxide. In most cases, an annealing at a temperature around 432°C under forming gas can remove the major part of these trapped charges in the gate oxide and therefore, returned the threshold voltage and the leakage currents almost to normal. The X-ray damages due to the metal evaporation are common for all evaporated metals but will depend on the evaporation temperature. It is obvious that the higher the evaporation temperature, the higher will be the dose of X-rays.

Unfortunately, gold has another drawback which prevents to perform an efficient annealing. It is well known that gold is not CMOS compatible due to its property to quickly diffuse (especially from temperatures around 330°C) into silicon or silicon oxide layers [160]. However, in our case, gold was deposited only on membranes, far enough from the electronics circuits, and furthermore, the chromium used as adhesion layer played a role of anti diffusion barrier. Nevertheless, it was too hazardous to risk an annealing at 432°C under forming gas. Our wafer covered with Gold was therefore annealed up to 320°C to avoid diffusion problems of gold into the silicon, during 30 minutes under nitrogen outside the clean-room to avoid contaminations problems. At the end, the threshold voltage as well as the high leakage currents remained drifted despite of the annealing which confirmed its poor efficiency.

Due to the addition of drawbacks of gold, we tried to replace it by another metal. **Tungsten** is CMOS compatible and has a high melting point but can be easily oxidized. Its lift-off showed not very accurate due to its high temperature of evaporation. Finally, we observed adhesion problems on oxide and bad selectivities in TMAH.

As proposed in a lot of groups working on gas sensors, we turned to **Platinum**, a standard in gas sensors [115]. It is currently used as microheater thanks to its high

3.2. INTERDIGITATED ELECTRODES: FROM DESIGN TO DEPOSITION

TCR but we prefered polysilicon to improve the CMOS compatibility and the thermal uniformity of our design. Pt is also widely used as gas sensing electrodes since it can not be oxidized in presence of air or humidity. Platinum needs to be deposited on an adhesion layer (like Cr, Ti or Ta) to adhere on silicon oxide but it has the advantage to allow direct wire bonding. It is fully CMOS compatible and allows therefore a complete annealing (at 432°C under forming gas) after its evaporation. Despite of its very high evaporation temperature (and therefore its higher production of X-rays in comparison with gold), the annealing after evaporation returned the electrical properties of our transistors almost to normal. See the next chapter for more details about our electrical characterizations.

Furthermore, the melting point of Platinum is very high which allows high working temperatures. Nevertheless, [111] and [161] reported that platinum is known to agglomerate at temperature above 700°C but higher temperatures are not necessary for our purpose. Its main drawback is its price, a little bit higher than gold. Finally, our tests revealed that it is inert in TMAH (pure or with additives).

About the adhesion layer, tantalum was used in [96], [99] and [116] (20 nm of Ta under 250 nm of Pt) on LPCVD nitride. [100][19][105][162][110] reported Cr (10 nm) used for adhesion of 300 nm of Pt on oxide. Titanium (20 to 50 nm) was proposed in [163], [97], [164], [118], [106] and [129] to replace chromium for platinum layers around 200 nm thick. Finally, [163] reported some degradations of Pt/Ti electrodes at very high temperature and [98] and [165] suggested to use TiN to avoid this problem.

Our first tests were performed using Ti as adhesion layer. Lift-off in acetone (+ ultra sound[1] to finish) was quick and easy to perform and adhesion on PECVD oxide was great. Nevertheless, titanium appeared underetched in TMAH (pure and with additives) after an etch of 1 hour and unsticked the platinum. New tests were achieved to replace titanium by chromium. But adhesion problems as well as damage in TMAH were also observed despite of pure chromium as well as pure platinum were tested inert in TMAH (see Section 1.6 in first part). Therefore, we decided to perform TMAH etching using the mechanical holder to protect the front side of the wafer during most of the etching period.

Finally, when heating the Pt-Ti electrodes up to 700°C during 15 minutes, cracks were observed on the electrodes (Fig. 3.3) such as those observed for gold. Nevertheless, the cracks observed had not the same nature than those observed in gold since the melting point of Platinum or Titanium is higher than the one of gold. Nevertheless, these cracks

---

[1]This common step to clean wafers in the industry does not provide any damage on CMOS circuits.

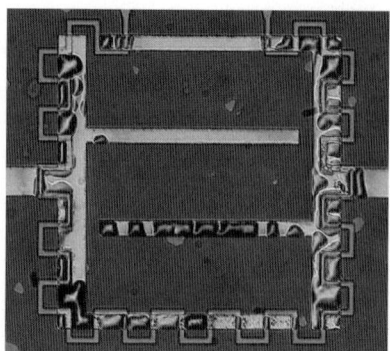

Figure 3.3: Cracks in interdigitated Pt-Ti electrodes when heated at around 700°C during 15 min (600x600 μm^2 membrane).

did not affect the resistivity measurements. An explanation of this behavior is analyzed in [161]. They reported that a 100 nm platinum film with a 10 nm titanium adhesion layer, degrades at temperatures greater than 700°C. Many mechanisms would appear to play a role in the degradation of the film, including the stress and interlayer reaction, but the dominant mechanism seems to be the agglomeration, a surface-diffusion-driven capillarity process. They proposed to increase the lifetime of the metallization by making the platinum thicker or using tantalum as the adhesion layer instead of titanium [161]. On the other hand, [98] proposed to replace tantalum by a sputtered TiN layer which is known to have a pronounced chemical inertness and does not react with Pt, even at very high temperature. Nevertheless, this layer needs to be sputtered and is therefore more expensive to deposit.

3.3 Sensitive layer deposition

Three kinds of sensitive layers were deposited on our interdigitated electrodes using different techniques. Well known **sputtering** is the more common and more reproducible technique but allows to deposit only very thin (around 0.6 μm) and dense layers. **Drop coating** and **screen-printing** techniques are alternative interesting techniques since they offer a higher thicknesses (between 2 and 20 μm for printing and around 10 μm for drop coating) and lower densities. Density and thickness fix the volumic surface in contact with gases and therefore the sensitivity to gases.

3.3. SENSITIVE LAYER DEPOSITION

Using a standard RF **sputtering** technique [166], sensitive layers of tin and tungsten oxides were deposited by lift-off on the membranes after the wafer backside etching, as the last step before the dicing and encapsulation. As described in previous chapters, the membranes yield was not damaged by the photolithography thanks to the high membrane robustness. The sputtered SnO_2 and WO_3 active layers were deposited using a reactive r.f. sputtering with the following parameters: work pressure $= 5*10^{-3}$ mbar, ambient temperature and gas composition made of O_2:Ar=1:1. After that, the lift-off was achieved in acetone and the whole wafers were annealed in air for 2 hours at 400°C.

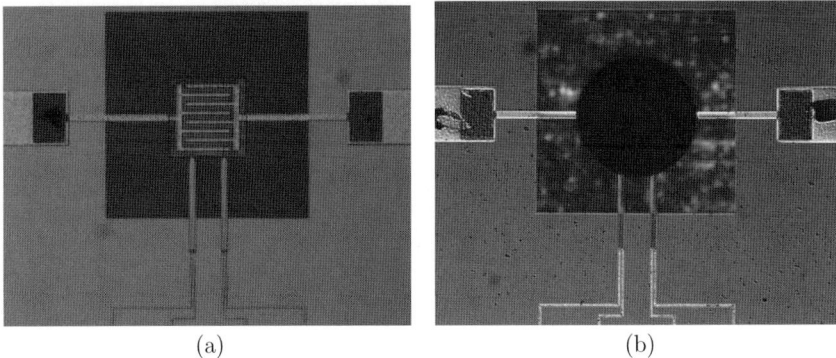

(a)   (b)

Figure 3.4: Picture of sputtered (a) and drop coated (b) sensors with WO_3 sensitive layer.

The **drop coating** was performed on other sensors after their packaging and wire bonding using an electronically controlled system formed by a pneumatic injection part connected to a syringe. This equipment allowed to deposit a drop of the paste over the membranes with desired parameters for a good reproducibility. The pastes were prepared from tin and tungsten nano-particles and an organic solvent. Once the deposition was made, the sensors were left for 10 minutes for leveling and after that were dried in an oven for 5 minutes at 150°C. This temperature was limited by the ceramic package used for our prototypes. The firing process was performed in situ using the sensors heating element at 400°C during 24 hours in order to eliminate completely the organic vehicle and to obtain adhesion of the active layer to the substrate.

The microstructure and composition of the sensing layers were investigated by scanning electron microscopy (SEM) as shown in Fig. 3.5 and energy-dispersive X-ray spectroscopy (EDX).

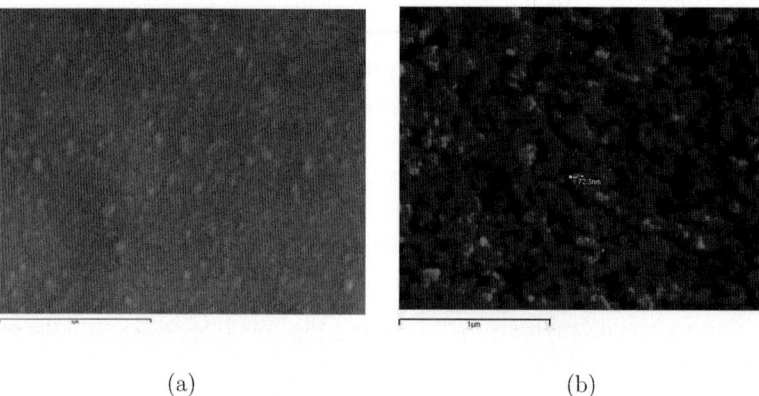

Figure 3.5: SEM micrograph of sputtered (a) and drop coated (b) SnO$_2$ sensor.

The thickness of the sensing layers was measured around 0.6 μm for the sputtered sensors (average particle size of 25 nm) and 10 μm for the drop coated layers (average particle size of around 60 nm) (Fig. 3.5(a) and (b)). The films showed an amorphous texture layer covering the substrate. The micrographs of the different layers showed that the deposited films are made by grains and voids. The voids within the film structure provide direct paths for gas molecules to flow in from the environment. An EDX analysis demonstrated that the used materials in the sensors active layer are tin and tungsten (the equipment was not capable to detect oxygen).

Sensitive layer deposition by **screen-printing** technique was also explored but did not give good results at this time due to non-dedicated equipment to perform correct alignments. We call screen-printing the transfer of a paste through a fabric screen onto a substrate [20]. This transfer occurs when the paste comes into contact with the substrate surface and is pulled through the screen. The high shear action of a squeegee passing over the screen allows this transfer to occur. The paste is deposited in a pattern that is defined by the open areas in the emulsion of the screen. For more details about screen-printing technique see [20].

Printing was performed before backside silicon etching and packaging to avoid damaging the sensor membrane with the high pressure needed to apply the paste [166][164]. The sensing layer consisted of a 5 μm thick SnO$_2$ nanopowder. The thickness of the active layer can be controlled between 2 and 20 μm after firing by adjusting paste viscosity, mask and printed thickness. The tin oxide was purchased from Sigma-Aldrich, Inc. (tin

(IV) oxide nanopowder (product ref. 54,965-7)). A printable paste was prepared by using an organic vehicle based on therpineol. After the tin oxide nanopowder was mixed with the vehicle, a paste with the required viscosity was obtained (filling was 58 %wt). The paste was printed onto the processed wafers by using a dedicated screen-printing machine that allows one-side mask alignment. To obtain a good adherence between the substrate and the active film, the wafer was heated to 60°C. After printing, the wafers were left for leveling at room temperature. They were subsequently dried for 15 minutes at 125°C for the organic vehicle to be completely dried, and then fired for 1 hour in a belt furnace at a single level of temperature, equal to 400°C. The deposited wafers were finally ready for backside etching in a mechanical holder to protect the front side surface with the sensitive printed layer from the etchant.

The sensitive layers of our first sensors were post-backed at 400°C to avoid irreversible drifts of the polysilicon heater as well as to avoid damaging of the gold interdigitated electrodes. Nevertheless, as introduced in the first chapter of this part, it would be better to perform the post-bake of the sensing layer at a temperature higher than the operating temperature in order to guarantee thermodynamic stability under the operating temperature, especially in case of screen-printed layers [128]. Furthermore, an annealing at a higher temperature would allow to use the gas sensor at a higher operation temperature which would increase its sensitivity to some gases (see our next results).

Our deep study on the operation of our microhotplate at high temperatures revealed that post-bake could be performed up to 700°C during 15 minutes. Directly after this annealing, the polysilicon microhotplate would work in a different way (with a lower room temperature resistance as well as a higher TCR) than before the bake. Nevertheless, after a dedicated ageing procedure of some days at the gas sensor operating temperature, the polysilicon microheater could return to normal.

3.4 Summary of the fabrication steps

We explained how the interdigitated electrodes were deposited and patterned on our SOI-CMOS processed wafers. Despite of the outlined ageing problems of the Pt-Ti layer at high temperature, we concluded that it seems to be the best option regarding CMOS compatibility, high operating temperature and fabrication easiness.

The three kinds of sensitive layers were then described and we explained how they were deposited on the interdigitated electrodes. We also revealed that depending on the deposition technique, the sensitive layer deposition is sequenced at different steps of the process. Sputtered layers are deposited at the end of the post-process, between the TMAH etching and the dicing and packaging. On the other hand, the drop coating is performed on the encapsulated chip, while the printing is done before the TMAH etching to avoid membrane breaks during the deposition.

Table 3.3 summarizes the common post-processing steps of the three kinds of gas sensors (sputtered, drop coated and screen-printed). Table 3.4 shows the following steps separately for each kind of gas sensors. All of these steps are fully compatible with the standard CMOS-SOI process performed before the first step of Table 3.3.

Table 3.3: Common post-processing steps of the three kinds of gas sensor.

	Steps	Thicknesses	Time
1	Photolithography for lift-off		
2	Ti + Pt evaporation	10 + 200 nm	
3	Lift-off in acetone + ultra-sound		13'
4	Bake under Forming Gas at 432°C		30'
5	Wafer thinning and polishing to 200 μm		
6	HF on backside to strip the native oxide		
7	Al evaporation on backside	500 nm	
8	Photolithography on backside		
9	Al etching in plasma on backside		7'
10	Photoresist stripping in fuming nitric acid		13'

Note that in step 5 of Table 3.3, the SOI processed wafers covered with their patterned metal layer were wafer back-side thinned and polished down to 200 μm. After thinning, a 500 nm aluminum layer was evaporated on back-side as protection layer for TMAH etching as well as back electrical contact for transistor measurements. This back side aluminum layer was then patterned to define the windows for the membrane release in TMAH.

Micromachining was performed in TMAH with additive not to damage aluminum. As described in the first chapter of this book, a 10 % TMAH solution at 90°C with 35 g/l of silicon powder and 15 g/l of APS was used to perform this step in 3 hours. Note that the same amount of APS was added a second time into the solution after 50 minutes of etching to compensate its dissolution. To avoid damaging the Pt-Ti interdigitated electrodes, the etching was performed in the holder. On the contrary, when chromium-gold was used instead of Pt-Ti as electrodes, the etching was done with the wafers fully immersed into

3.5. MEASUREMENTS RESULTS WITHOUT GAS

Table 3.4: Following post-processing steps separately for each kind of gas sensor.

	Sputtered	*Drop coated*	*Screen-printed*
11			**Screen-printing**
12	TMAH 10 % etching with Si and APS at 90°C (in the holder)		
13	Rinsing DI water (hot and cold) + methanol		
14	Metal oxide layer **sputtering**		
15	Final bake under Forming Gas at 432°C		
16	Dicing and packaging		
17		**Drop coating**	

the solution. Wafers were then rinsed in boiling DI water followed by a second rinsing in cold DI water and finally dryed in methanol. As explained in the next chapter, the final bake under forming gas finally removed the possible interface traps which could appear on backside after the TMAH etching and which could affect the characteristics of some integrated devices.

3.5 Measurements results without gas

Electrical measurements were fully performed on microheaters and interdigitated electrodes without gas. First of all, power consumption before and after coating were compared as shown in Fig. 3.6. A slight power increase of around 5 mW at 400°C was observed after drop coating. This increase is due to the higher thermal conductivity of the metal oxide layer. Same results were more or less observed on sputtered sensors despite of their thinner thicknesses.

Gas sensing resistance measurements versus time revealed that the gas sensing layers need some time to be stabilized. Both constant current and constant voltage measurements were achieved and compared. Fig. 3.7(a) shows the microheater resistance and the gas sensing layer resistance relaxation as a function of time when applying a constant voltage on the microheater corresponding to a starting current value of 1.4 mA. The gas sensitive layer in this case was a sputtered WO_3 layer measured by 8 interdigitated fingers pairs. Fig. 3.7(b) compares the gas sensing resistance relaxation versus time when constant current or constant voltage was applied on the microheater. The layers as well as the amount of fingers were the same in each measured sensors. This interesting comparison revealed that the measurement at constant voltage allows a quicker stabilization of the sensing layer resistance. Furthermore, both methods seemed to reach the same resistance value at the end.

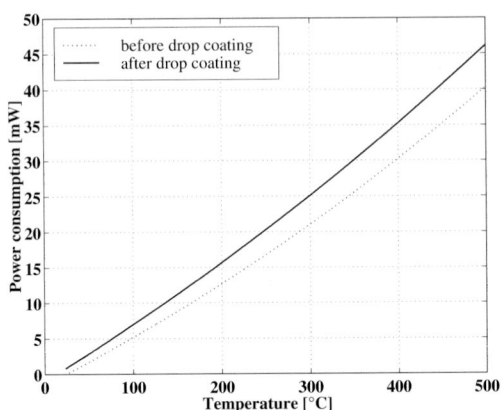

Figure 3.6: Power consumption versus temperature: comparison in presence and without WO_3 drop coated layer on a 600x600 μm^2 membrane with 4 pairs of interdigitated fingers.

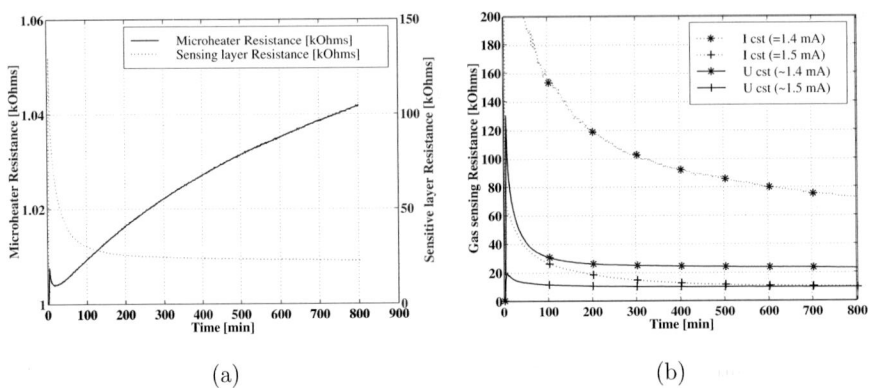

Figure 3.7: (a) Microheater resistance and gas sensing layer resistance relaxation versus time (WO_3, 8 fingers pairs, constant voltage corresponding to a current of 1.4 mA, 600x600 μm^2 membrane); (b) gas sensing resistance relaxation comparison when constant current or constant voltage is applied on the microheater (idem).

3.6. MEASUREMENT RESULTS UNDER GAS AND DISCUSSIONS

Our measurements without gas on different interdigitated fingers designs revealed that the measured resistance values depended more on the temperature applied by the microheater than on the amount of fingers pairs. So in our case, stabilized resistance values between 20 and 40 kΩ were commonly reached.

Finally, accelerated ageing tests were started on gas sensors in order to complete our realibility tests performed on the membrane covered by the microhotplate only. While increasing the power applied on the microheater up to the membrane break down, mechanical deformations, electrical measurements and optical observations were used to characterize the ageing process. Unfortunately, technological problems on our interferometric test bench did not allow us to draw conclusions at this time. Same tests were reported in [116] (with a tantalum-platinum microhotplate) and revealed that the interdigitated electrodes haved a slight effect on the maximum power before failure and on the membrane deformation, but haved a larger impact on the microhotplate lifetime. On the other hand, the addition of the gas sensitive coating would improve the mechanical robustness of the membrane [116].

3.6 Measurement results under gas and discussions

To enable gas-sensing measurements with the fabricated sensors, a test chamber with volume of 125 cm^3, which permits the independent power control of the installed sensors, was used. Measurements were performed by applying a constant voltage (adjusted to the temperature desired value) on the microheater during each gas acquisition. Using an RS-232 interface and a data acquisition system HP 34970A, the measured values of the resistance were transferred to the PC, where written-in-house MATLAB program acquired in real time the resistance and working temperature of the sensor. A series of tests using two different types of measurement systems were performed.

The first of them was a continuous flow system (Fig. 3.8(a)). All contaminants were introduced into the test chamber using dry air as carrier gas. The gases under test with this system were carbon monoxide (CO), nitrogen dioxide (NO$_2$) and methane (NH$_3$). During the measurements with this system, the relative humidity was kept between 10 and 15 %. All the measurements were repeated three times in order to verify the obtained results. Before the initial measurement at some temperature the sensors were left for 24 hours in order to obtain stable values, while between every two measurements a pause of twenty minutes was made.

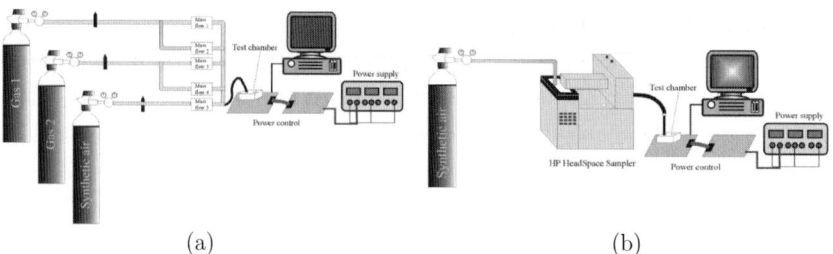

(a) (b)

Figure 3.8: Measurement system based on a (a) continuous flow, which delivers constant concentrations of gases or gases mixtures to the sensor chamber using a dry air as carrier gas; (b) static headspace auto-sampler.

In the second set of experiments we used a HP Headspace Auto-sampler to inject the vapors of the tested species into the chamber (Fig. 3.8(b)). At the beginning of the measurement, the system waited for 75 seconds and subsequently injected a sample of the tested volatiles in the airflow. This system allowed tests under ethanol, acetone and ammonia vapors. The sensors detected the volatiles (we can observe the sharp decrease in sensor resistance represented in Fig. 3.9) and soon after the exposure, their resistance increased towards the baseline value.

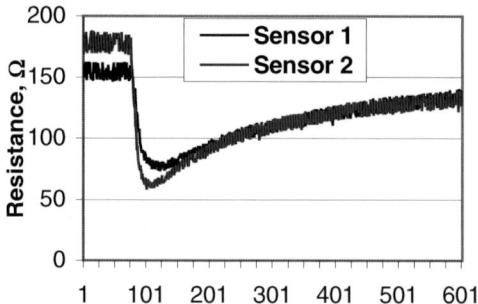

Figure 3.9: Response to ethanol of the SnO_2 drop-coated sensors.

This reveals that after the volatiles had crossed the sensor chamber, the sensors were in the presence of clean air again. Every measurement took 10 minutes to complete. This time was enough in order to obtain the same value for the initial resistance of the tested sensors.

The sensitivity of 4 micro-sensors was studied: 2 sputtered during the first measure-

3.6. MEASUREMENT RESULTS UNDER GAS AND DISCUSSIONS

ment cycle and 2 drop coated during the second measurement cycle. Sensors were operated at three different temperatures (150, 200 and 250°C) set by applying the dedicated voltage to their heating resistor. The sensitivity of the sensors has been defined as R_a/R_g; where R_a is the resistance in dry air and R_g is the resistance in presence of gas. The results show that the sensors sensitivity increases with the operating temperature. The tested sensors showed high selectivity to ethanol, ammonia and NO_2.

Both type of active layers (SnO_2 and WO_3) deposited with either r.f. sputtering or drop-coating, were sensitive to ethanol vapors (Fig. 3.10).

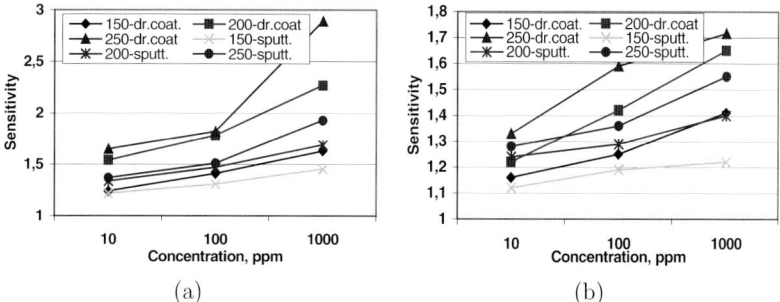

Figure 3.10: Average sensitivity of the SnO_2 (a) and WO_3 (b) sensors to ethanol, at different operating temperatures.

At the same time, the drop-coated WO_3 sensors, showed good response to ammonia and NO_2 (Fig. 3.11).

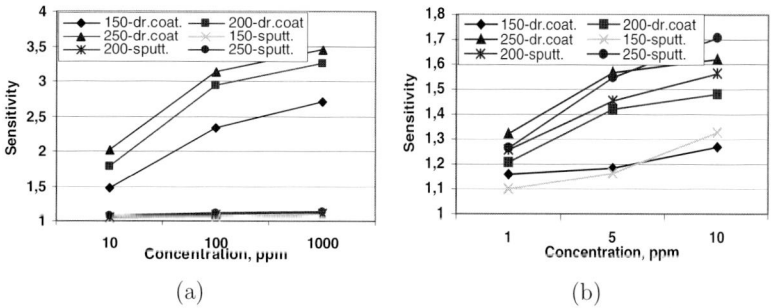

Figure 3.11: Average sensitivity of the WO_3 sensors to (a) ammonia and (b), NO_2, at different operating temperature.

The response to NH_3 and NO_2 of the WO_3 sensors was expected, because of the high affinity of this active material to the nitrogen species [167][168][169]. None of the sensors responded to acetone, methane and CO which could be explained by the low operating temperature.

The sensor response time ranged between 30 and 45 s, depending on its operating temperature. It was defined as the time it took to reach 90 % of its steady value after exposure to a gas. Generally, the sensor response was faster when operated at higher temperature.

The response of the sensors to variations in the ambient moisture was also studied. It was observed that the sensor resistance decreased when the relative humidity was increased (Fig. 3.12). Furthermore, when the sensors were operated at higher temperatures, the change in their resistance caused by a change in the moisture level was higher.

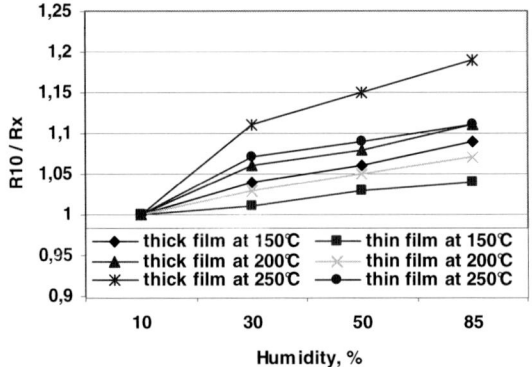

Figure 3.12: Response of the drop-coated and sputtered sensors to humidity changes at different operating temperatures. In the legend, thick designates the drop-coated sensors, and thin the sputtered ones.

Fig. 3.12 shows the ratio between the resistance of the sensors at 10 % r.h. (this humidity level is taken as a reference value) and the values for different humidity levels (up to 85 % r.h.). From Fig. 3.12, it can be derived that the devices with drop-coated active layers are more affected by changes in the moisture level than the sputtered ones. These results are in accordance with previous studies on gas sensors with thin and thick active layers [170]. However, the response to moisture can be considered as moderate, when compared to the sensitivities found for the different vapors studied.

3.6. MEASUREMENT RESULTS UNDER GAS AND DISCUSSIONS

2 fingers (54 µm) membrane 600*600 µm²

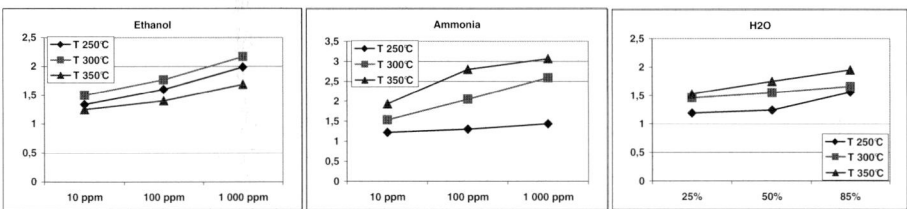

4 fingers (18 µm) membrane 600*600 µm²

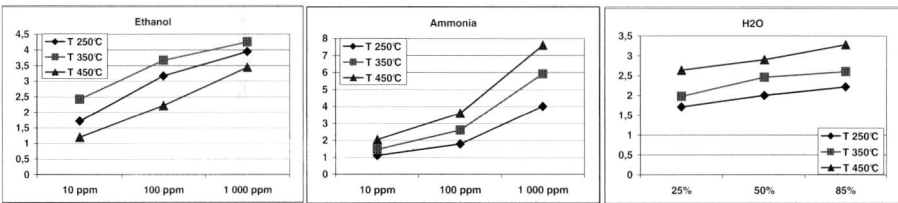

8 fingers (6 µm) membrane 600*600 µm²

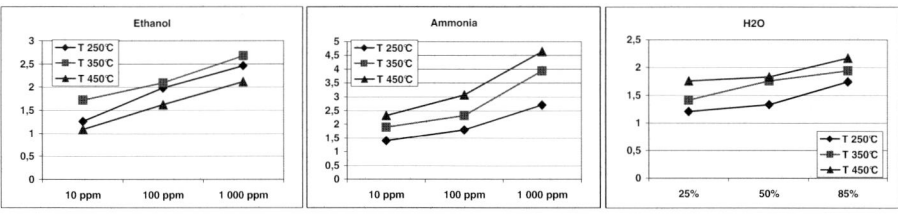

Figure 3.13: Results obtained on sputtered gas sensors with the new design of interdigitated electrodes (2, 4 and 8 fingers pairs) under ethanol, ammonia and humidity.

All of the results presented above were obtained with the old design of interdigitated electrodes (with 17 fingers). Results of our measurements on WO_3 sputtered gas sensors with the new design of interdigitated electrodes have been obtained more recently and proved better than with the old design. All of these results are summarized in Fig. 3.13 under ethanol, ammonia and humidity for different temperatures and interdigitated electrodes designs. Note that these new sensors did not respond to methane despite of the higher operating temperature than in our previous measurements.

3.7 Conclusions

The design and all the fabrication steps necessary to transform our microhotplate into a whole SOI-CMOS compatible gas sensor were discussed. Our tests with different materials usable as interdigitated electrodes revealed that gold needed to be advantageously replaced by Platinum (on Titanium or chromium as adhesion layer) in order to provide CMOS compatibility as well as fabrication simplicity and to allow high operating temperature. Gold had nevertheless the advantage to stand fully immerssion into the TMAH bath without damage. Unfortunately, we observed that it was not the case anymore with platinum on titanium (or chromium) due to adhesion damage during the etching. Improvements in deposition could be done in order to increase in the future the adhesion between platinum and titanium (or chromium). Nevertheless, the use of the mechanical holder at this time allowed to perform efficient etching without damage of the front surface.

Tin dioxide and tungsten trioxide sensors have been fabricated using either reactive r.f. magnetron sputtering or drop-coating. The fabricated gas sensors were tested to ethanol, acetone, ammonia, carbon monoxide, nitrogen dioxide and methane, showing promising results and excellent selectivity to ethanol, ammonia and NO_2. The sensing parameters, such as response, recovery time and the impact of the humidity on the sensor response have also been examined. As the results from the drop-coated sensors were better, gas sensitivity improvements are in progress at this time using screen-printed active layer, because better results can be obtained with thicker and more porous films.

In conclusion, we demonstrated that the use of completely CMOS compatible micro-machining techniques during the fabrication process, allowed to perform smart gas sensor co-integrated with their electronics in SOI-CMOS technology.

Chapter 4
SOI-CMOS compatibility validation

4.1 Introduction

In order to validate the full SOI-CMOS compatibility of our sensors process, semiconductor devices were integrated close to the sensors and measured after each step of the post-processing.

Transistors of different sizes, different types (N and P) and different doping levels were integrated as well as MOS capacitors and gated diodes in SOI technology (Fig. 4.1 and 4.2). These devices were firstly fabricated using a standard CMOS SOI process with its additional intermediate processing steps to build the second constitutive layer of the membrane (see its whole description in Chapter 1 of this part). The full compatibility of these additive steps was confirmed by comparing our results with measurements on wafers whiwh followed a completely standard process. It was expected since the additional steps were performed far enough from the devices.

First of all, this chapter starts with a summary of the basics of SOI technology, the main effects which may affect the electrical properties of devices and the properties of oxide and Si/SiO_2 interface traps. It will facilitate the understanding of the measurements discussion. The post-processing steps are then summarized, followed by the description and the discussion of our experiments.

A final demonstrator is then described consisting in the integration of transistors on membrane. It demonstrates in one device the full compatibility of our whole process and post-processing as well as the co-integration of the sensors fabrication steps and the standard IC processing.

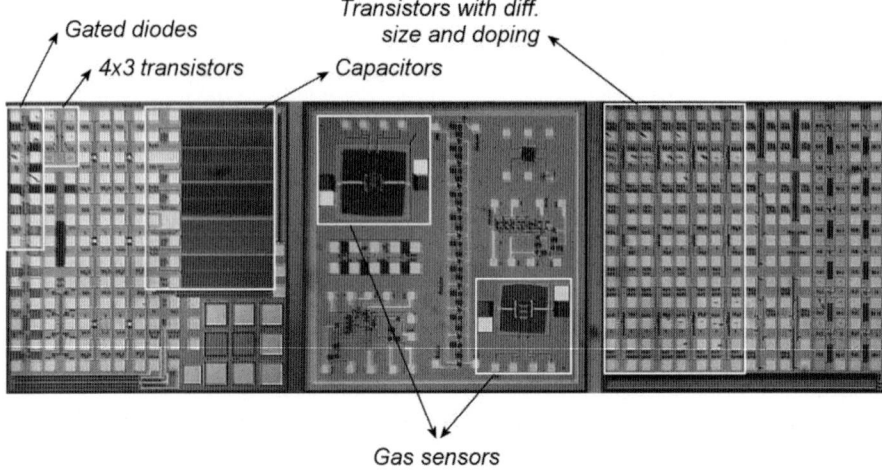

Figure 4.1: Co-integration of transistors of different sizes, different types (N and P) and different doping levels, MOS capacitors, gated diodes and gas sensors in SOI CMOS technology.

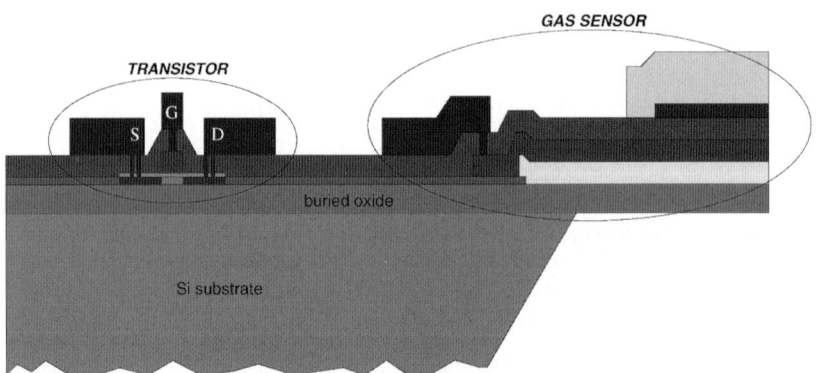

Figure 4.2: Cross-section of a gas sensor co-integrated with a transistor.

4.2 Basics of SOI technology

By isolating a thin silicon layer from the substrate via a buried silicon oxide, SOI (Silicon-on-Insulator) technology features numerous advantages over conventional bulk silicon (Fig. 4.3), such as full dielectric isolation of the devices; absence of latch up; reduced parasitic capacitances and short-channel effects; higher speed; easier CMOS processing and MEMS co-integration, which makes SOI the first candidate for sub-micron low-voltage, high speed, low-power co-integrated circuits.

At present, the ideal SOI materials candidates for VLSI CMOS applications are SIMOX (Separation by Implanted Oxygen) and UNIBOND materials. The principle of the first one consists in the formation of a buried layer of SiO_2 by implantation of oxygen ions beneath the surface of a silicon wafer. The second one is produced by the so-called Smart-Cut process. It combines hydrogen ion implantation technology and wafer bonding technique to transfer a thin surface layer from a wafer onto another wafer or an insulating substrate [171].

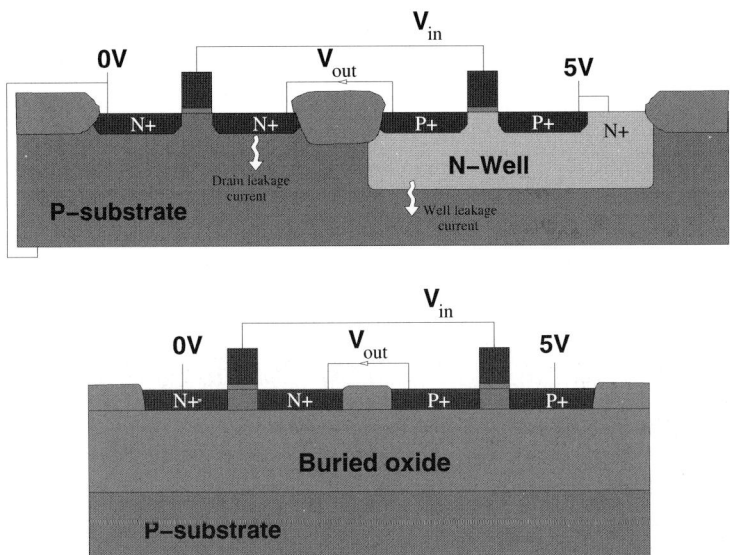

Figure 4.3: Cross sections of a bulk (upper part) and a SOI (lower part) CMOS inverter.

According to the thickness of the top silicon film, SOI MOSFETs can be classified into two different structures: thick-film partially-depleted (PD) and thin-film fully-depleted

(FD) SOI devices. In PD operation, there is no interaction between the depletion zones arising from the front and the back interfaces, so that a neutral region remains beneath the front depleted zone. When this neutral body is left electrically floating, two parasitic effects may arise: one is called the "kink effect" which leads to a degradation of the output conductance in saturation and the other is a parasitic open-base NPN bipolar transistor between source and drain which may lead to early breakdown [171].

In contrast to thick film devices, the thin-film FD ones are virtually free of the kink effect thanks to the full depletion of the silicon film. Furthermore, this kind of SOI devices exhibits the most attractive properties such as low electric fields, high transconductance, excellent short-channel behaviour and a quasi-ideal subthreshold slope. Therefore, thin-film SOI MOSFETs can be considered as the most promising structure for sub-micron integrated circuits.

4.2.1 The main SOI technology issues

In the following section, we summarize the main issues generally observed in SOI devices [172][173][174][175][8] which will be referred to further in the text. These effects mainly appear in partially depleted devices. Nevertheless, under particular electrical or technological conditions, some floating-body effects can also be observed in fully depleted devices or in devices designed fully depleted but which actually appear just nearly fully depleted. Such behavior was observed in our case under given doping conditions. The fully or partially depleted behavior is indeed dependent on the respective thicknesses of the depletion region and of the silicon film. It is well known that the depletion region depth is inversely proportional to the channel doping level. Therefore, if the doping level is higher than expected, a transistor could be in a regime intermediate between partially and fully depletion.

In a floating-body partially depleted or nearly fully depleted SOI n-MOSFET driven far enough in saturation, a parasitic behavior called **kink effect** can appear. The drain current abruptly rises which leads to a conductance discontinuity. The origin of this parasitic effect lies in the impact ionization of high energy electrons, the so-called "hot electrons" which creates supplementary electron-hole pairs below the gate. Hot carriers can transfer enough energy to other electrons to be injected through the gate oxide and lead to a gate current while the holes are attracted towards the lowest potential. Nevertheless, since the buried oxide prevents these holes from being evacuated to the substrate,

4.2. BASICS OF SOI TECHNOLOGY

as in a standard bulk device, these holes accumulate in the floating body and raise its effective potential. Consequently, the threshold voltage is lowered and the drain current increases faster. As the impact ionization is more severe for n-channel MOSFETs than for p-channel, the kink effect is more pronounced in nMOS transistors.

To avoid the kink effect in PD transistors, the body needs to be connected by a lateral contact (of P+-type in case of an enhancement-mode nMOSFET). In this case, the partially depleted SOI transistor follows the behavior of a transistor in bulk technology since the extra holes created by the impact ionization are sucked out via the tied body contact. On the contrary, when the transistors are fully depleted with a thin enough silicon film, the kink effect vanishes since hole accumulation becomes inefficient [171].

It must be noted that the gate current can be also due to **direct tunneling** or **Fowler-Nordheim tunneling**. Nevertheless, the direct tunneling can usually be neglected for gate oxide having thicknesses higher than 7 nm. For more information on this subject, see works of [176] and [177].

The **substrate** or **body effect** can be defined as the dependence of the threshold voltage on the back-gate bias. In a bulk transistor, when a negative bias is applied to the substrate or to the well surrounding the device (with respect to the source), the threshold voltage increases as a square-root function of the substrate or well bias. In a thick-film SOI (partially depleted), the substrate effect can be neglected because there is no coupling between front and back gates. However, a body effect can exist as in bulk when the PD body is contacted and biased. In thin-film FD SOI devices, the body effect can be calculated as the derivative of the front threshold voltage according to the back gate voltage since the front and back gates are couplet through the FD Si film. It should be noted that this effect is nevertheless much smaller than in a bulk MOSFET resulting in more ideal characteristics.

An other effect called the **self-heating** appears at high drain current levels in SOI devices. SOI transistors are indeed thermally insulated from the substrate by the buried insulator. As a result, removal of excess heat generated by Joule effect within the device is less efficient than in bulk, which yields to substantial elevation of device temperature [171]. This leads to a reduction of the carrier mobility and hence of the drain current level. The effect can appear in $I_d - V_d$ characteristics as a negative resistance. High temperatures also reduce the impact ionization rate and thus, in a floating body SOI MOSFET, will change the body potential and reduce the kink effect. This effect will be especially observed in SOI devices on membrane due to their higher thermal insulation.

4.2.2 The transconductance over drain ratio current

In the literature [178][179], $I_d - V_g$ and $I_d - V_d$ characteristics are commonly extracted alone for a complete check of the compatibility of a process. Other characteristics such as the transconductance over drain current ratio g_m/I_d versus drain current normalized to width/length aspect ratio $(I_d/(W/L))$ reveals further interesting information on devices.

The relevance of this data derives from 2 main reasons [180]. Being size independent, it is then a unique characteristic for all transistors of the same type (NMOS or PMOS) and doping level in a given process. To the first order, it is then just dependent on the device operating region. It can be approximated by a continuous MOS model (such as the EKV). In strong inversion, it yields valuable assessments about the linearized body effect factor n and the effective mobility μ in the channel as confirmed by Eq. 4.1,

$$\frac{g_m}{I_d} = \sqrt{\frac{2\mu C_{ox} W/L}{n I_d}} \qquad (4.1)$$

where Cox is the gate oxide capacitance, W the width of the transistor and L, its length [171]. The **body effect factor** n represents the image of the efficiency of the coupling between the gate voltage and the channel. It can be expressed as the ratio of the capacitance between the channel and the back gate, to the capacitance between the gate and the channel. The first capacitance reprensents the "force" that tends to prevent the potential in the channel from following gate voltage variations [171].

4.2.3 The properties of the oxide and the Si/SiO$_2$ interface traps

It is worth to remind that the SiO$_2$ and the Si/SiO$_2$ interface both contain various charges and traps which will play a significant role in the study of this chapter. Their nature and amount depend on the process steps. Some are created by the standard process itself; others supplied during the post-processing and need to be well understood since these charges can be responsible for threshold voltage shifts. Four types of charges are known to exist in the oxide or near the Si/SiO$_2$ interface [181].

Fixed oxide charge Q_f are located within 35 Å of the Si/SiO$_2$ interface in the so-called transition region between silicon and SiO$_2$. These do not change their charge state by exchange of mobile carriers with the silicon. The Q_f charge centers are predominately positive and lead to shift the V_t to more negative values.

4.2. BASICS OF SOI TECHNOLOGY

Mobile ionic charge, Q_m, are commonly caused by the presence of ionized alkali metal atoms (like Na^+ or K^+). The amount of mobile charge incorporated into the SiO_2 depends mainly on the cleanliness of the oxidation process. Major sources of Na during processing are: gate or contact metallization, oxidation and annealing furnaces and gases, photoresist bake and incomplete resist stripping and contaminated chemicals. This type of charge is located either at the gate/SiO_2 interface or at the Si/SiO_2 interface, to where it drifts under the presence of an applied positive field to the gate. This drift across the oxide can occur even at room temperature, since these ions are extremely mobile in SiO_2. The presence of mobile charge can also result in long term changes in the device threshold voltage, as the ions drift from the gate to the Si/SiO_2 interface.

The **interface charge trap** Q_{it} refers to charge which is localized at the Si/SiO_2 interface on sites that can change their charge state by exchange of mobile carriers (electron or holes) with the silicon. Q_{it} depends mainly on the oxidation parameters. These charge can be reduced with a low temperature post-metallization anneal at e.g. 432°C in forming gas (10 % H_2 - 90 % N_2) for up to 30 min. Aluminum must be present for the anneal to be effective. It is believed that water in SiO_2, present even in dry oxides, reacts with the aluminum to form Al_2O_3 and atomic hydrogen (H). The H diffuses to the Si/SiO_2 interface, where it chemically reacts with traps, rendering them electrically inactive. Interface charge traps can also appear under radiation exposure as it is the case during e-beam evaporation [182].

Oxide trapped charge Q_{ot} can be located at the gate/SiO_2 interface, the Si/SiO_2 interface, as well as deep in the oxide. The traps are associated with defects in the SiO_2, such as impurities and broken bonds. They can be reduced by annealing conditions similar to those established to reduce Q_{it}. Such traps are usually uncharged, but can become charged when electrons and holes are introduced into the oxide and trapped at the trap site. One way that traps become charged is by avalanche injection of highly energetic hot carriers (electrons or holes) into the oxide. Exposing the oxide to radiation environments can also produce significant levels of trapped charges. The process itself like e-beam evaporation generating X-rays, sputtering which generates UV-rays or plasma etching steps [179] can result in radiation damage to silicon devices [182].

In SOI technology, the presence of these charges and interface traps may be more difficult to understand since the thick buried oxide increases the probable sites of traps.

4.3 Post-processing steps

A the end of the standard SOI CMOS process with its few additional steps to build the second constitutive layer of the membrane, the following steps were performed on our wafers as summarized in Table 4.1.

Table 4.1: Post-processing steps on the fully processed CMOS-SOI wafer.

	Steps	Thicknesses	Time
1	**Electrical measurements before evaporation**		
2	Photolithography for lift-off		
3	Ti + Pt evaporation	10 + 200 nm	
4	Lift-off in acetone + ultra-sound		13'
5	Bake under Forming Gas at 432°C		30'
6	**Electrical measurements after evaporation**		
7	Wafer thinning and polishing to 200 μm		
8	HF on backside to strip the native oxide		
9	Al evaporation on backside	500 nm	
10	Photolithography on backside		
11	Al etching in plasma on backside		7'
12	Photoresist stripping in fuming nitric acid		13'
13	**Electrical measurements before TMAH**		
14	TMAH 10 % etching with Si and APS at 90°C		30'
15	Rinsing DI water (hot and cold) + methanol		
16	**Electrical measurements after TMAH**		
17	Final bake under Forming Gas at 432°C		30'
18	**Electrical measurements after final bake**		

The lift-off of Titanium and Platinum was dedicated to the interdigitated electrodes for gas sensing. The aluminum layer on backside provides not only the back gate contact but also a mask for the TMAH etching. As for the two bakes under forming gas, their reason will be explained further.

4.4 Measurements

Measurements were carried out on transistors of different gate sizes (20x20 μm, 3x3 μm and 4x3 μm), different type (N and P) and 4 different channel doping levels. They were performed after each critical step of the post-processing as shown in Table 4.1. Table 4.2 summarizes the doping parameters of the 4 different channels as well as the doping level

4.4. MEASUREMENTS

of the source and the drain. Note that the first type "**I**" denotes intrinsic silicon without intentional doping and that "**P12**", a combination of P1 and P2 doping levels.

Table 4.2: Doping levels of the 4 different channels and of the source and the drain.

Type		Doping source	Energy [KeV]	Dose [cm^{-2}]
CHANNEL	I	x	x	x
	P1	Bore	18	2.3×10^{11}
	P2	Bore	18	1.3×10^{12}
	P12	Bore	18	P1 + P2
SOURCE	N	Arsenic	40	4.5×10^{15}
& DRAIN	P	Bore	20	5.6×10^{15}

The silicon substrate as well as the top silicon film are initially doped at a concentration of about 1×10^{15} cm^{-3} which therefore fixes the doping level of the intrinsic channel. Used wafers were UNIBOND SOI 3" wafers. The final layer thicknesses after full process were:

- top silicon film: 80 nm;

- gate thermal oxide: 31 nm;

- buried oxide: 400 nm.

$I_d - V_g$ characteristics with back gate fixed at 0 V, $I_d - V_g$ with variable voltage on the back gate and $I_d - V_d$ characteristics were extracted as well as the gm/I_d versus the drain current normalized to width/length aspect ratio ($I_d/(W/L)$) in order to explain the possible variations from one step of the post-processing to an other. Measurements were also done on large MOS capacitors to extract the $C - V$ characteristics and the gate current. Finally, we will report measurements of gated diodes to illustrate the recombination currents in the silicon volume as well as in the top and back interfaces after each step of our post-process.

All the measurement results we present are representative of the behavior of several identical devices distributed over the same wafer.

4.4.1 MOS transistors

Measurements results

As a first observation, our measurements on MOS transistors after the evaporation revealed a threshold voltage shift to the left which strongly suggests that charges are trapped in the gate oxide [183]. The shifts appeared proportional to the transistor size and slightly different from one doping level to an other. The other post-processing steps did not provide further drifts. Fig 4.4 reports the $I_d - V_g$ characteristics during the post-processing steps (except the final post-bake) for two sizes in all cases (N and P types and the four channel doping levels). Note that the reported threshold voltage difference values were measured at a drain voltage of 100 mV using the maximum of the second derivative of I_d with respect to V_g as method. The difference was measured as the threshold value after the TMAH etching minus the measured threshold value before evaporation (at the beginning of the post-processing).

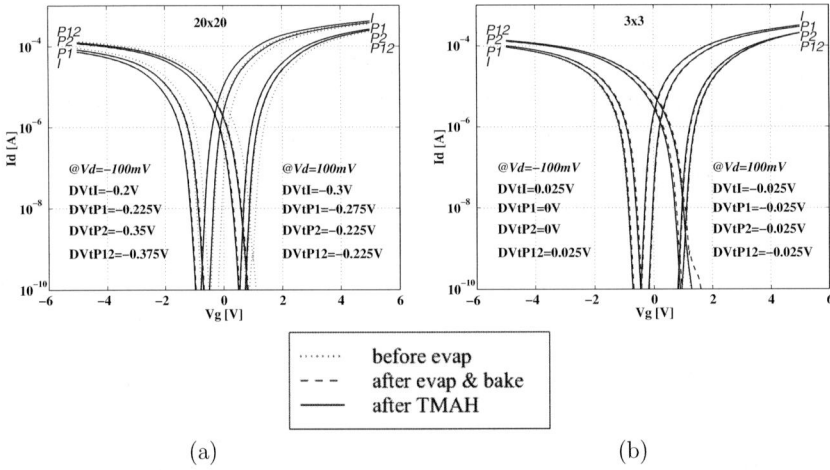

Figure 4.4: $I_d - V_g$ measured characteristics during the post-processing steps for nMOS and pMOS transistors with channel doped I, P1, P2 and P12. The back gate was fixed at 0 V and V_d=1.55 V for nMOS and V_d=-1.55 V for pMOS. (a) transistors size 20x20 μm^2 and (b) transistors size 3x3 μm^2.

The final post-bake under forming gas revealed a slight recovering of the 20 μm x 20 μm transistors appearing as a slight drift of the $I_d - V_g$ curves measured after the TMAH etching to the right as illustrated in Fig. 4.5 for nMOS and pMOS transistors with a

4.4. MEASUREMENTS

channel type P1. This shift appeared negligible in the 3x3 μm^2 transistors as we can observe in Fig. 4.6.

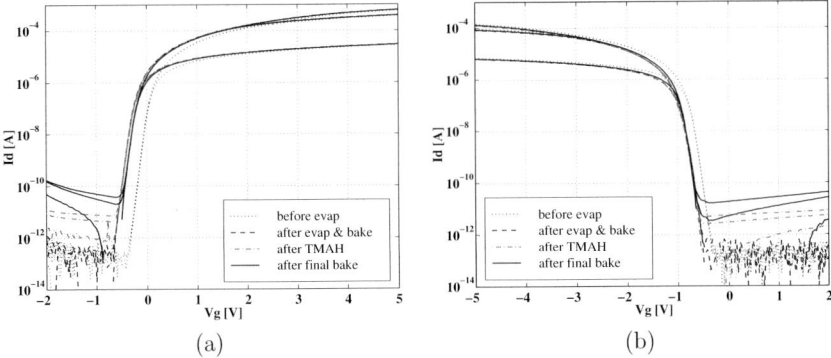

Figure 4.5: $I_d - V_g$ measured characteristics during the post-processing steps for 20x20 μm^2 transistors (channel doped P1) (a) nMOS with V_d=0.1; 1.55 and 3 V and (b) pMOS with V_d=-0.1; -1.55 and -3 V. The back gate was fixed at 0 V.

Fig. 4.5 and 4.6 also reveal a slight increase of the leakage current after the TMAH etching in both transistor sizes (20x20 μm^2 in Fig. 4.5 and 3x3 μm^2 in Fig. 4.6). The leakage current returned to usual values after the final bake in the smaller transistors (3x3 μm^2) but increased more in the larger ones (20x20 μm^2).

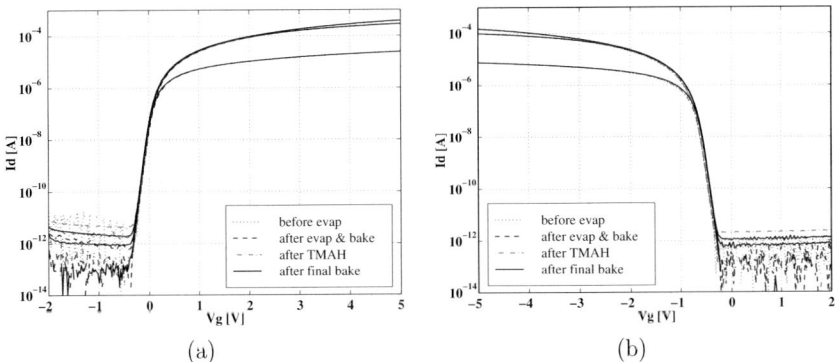

Figure 4.6: $I_d - V_g$ measured characteristics during the post-processing steps for 3x3 μm^2 transistors (channel doped P1) (a) nMOS with V_d=0.1; 1.55 and 3 V and (b) pMOS with V_d=-0.1; -1.55 and -3 V. The back gate was fixed at 0 V.

The transconductance over drain current characteristic g_m/I_d versus drain current normalized to width/length aspect ratio ($I_d/(W/L)$) revealed interesting informations. As it can be observed in Fig. 4.7 in strong inversion ($g_m/I_d < 10$ V^{-1}), the g_m/I_d decreased after evaporation and was slightly recovered after the final bake for the nMOS transistors (Fig. 4.7(a)). For the pMOS (Fig. 4.7(b)), the g_m/I_d decreased after evaporation and showed a higher value than its initial value after the final bake. This behavior can be explained by a decrease of the mobility and a higher body effect factor for the n-transistors and a higher mobility and a body effect decrease for the p-transistors. On the contrary, we did not observe any shift of the g_m/I_d curves on the 3x3 μm^2 transistors during the post-processing.

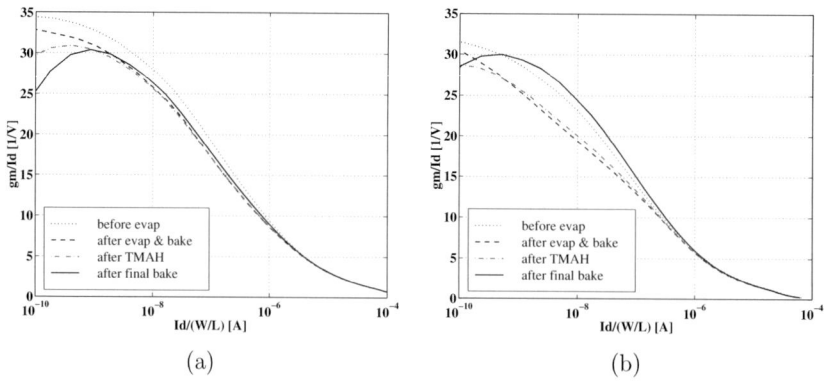

Figure 4.7: $g_m/I_d - I_d/(W/L)$ measured characteristics during the post-processing steps for 20x20 μm^2 transistors with channel doped P1 (a) nMOS with $V_d = 1.55$ V and (b) pMOS with $V_d = -1.55$ V. The back gate was fixed at 0 V.

As can be observed in Fig. 4.4(a), if the doping concentration is increased (to P2 and P12 instead of P1) in the 20x20 μm^2 transistors, the threshold voltage drift decreases for the N type transistors but increases for the P ones. Due to their too high doping level respective to the film thickness, the P2 and P12 doped channels of the N type transistors appeared to be nearly fully depleted. This is observed on the subthreshold slope variation when the drain voltage V_d was increased up to 3 V which shows a floating body effect. It was confirmed again during measurements with variable back gate voltage which do show the usual FD front threshold variation. The effect is worse in 3x3 and 4x3 μm^2 transistors (see Fig. 4.9(a)); nevertheless, it can not be observed by comparison of their threshold voltage drifts due to their too low values (see Fig. 4.4(b)). On the contrary, the P type transistors were fully depleted for all channel dopings. Note that our transistors were designed to work in fully depletion for all doping levels, but the unexpected nearly fully

4.4. MEASUREMENTS

depleted behaviors due to process variations will allow us to draw easier conclusions about the location and the repartition of the charges in the oxide layers and their interfaces.

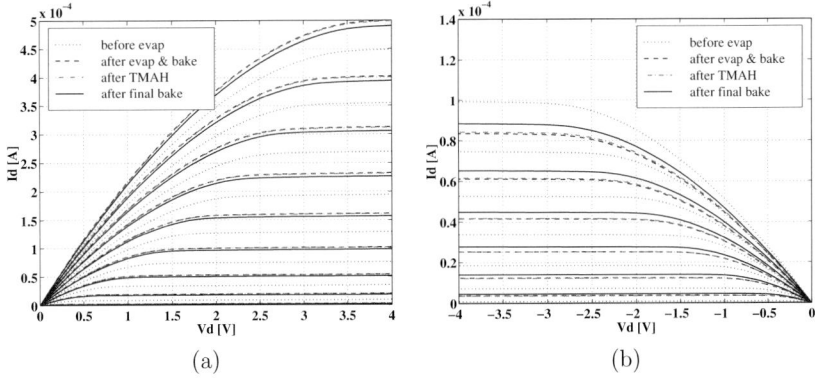

Figure 4.8: $I_d - V_d$ measured characteristics during the post-processing steps for 20x20 μm^2 transistors with channel doped P1 (a) nMOS transistors with V_g varying from 0 to 4 V by steps of 0.5 V and (b) pMOS transistors with V_g varying from 0 to -4 V by steps of -0.5 V. The back gate was fixed at 0 V.

Finally, the $I_d - V_d$ curves extracted from the 20x20 μm^2 transistors (with channel doped P1) confirmed our observations reported on the g_m/I_d curves as we can see in Fig. 4.8. Furthermore, a slight kink effect was observed on nMOS transistors doped P2 and P12 which confirms that these transistors were nearly fully depleted.

To conclude, Fig. 4.9 summarizes our results for N and P transistor types featuring a representative gate size of 4x3 μm^2, and the standard doping levels P2 (for type N) and P1 (for type P). It confirms that the N type transistor is nearly fully depleted at these doping levels as we had observed from the steeper subthreshold slope when the drain voltage reaches 3 V (Fig. 4.9(a)) and the kink effect appearing in the $I_d - V_d$ curves (Fig. 4.9(e)). Nevertheless, N type and P type transistors revealed threshold voltage drifts (measured at 100 mV) of respectively -50 mV and -25 mV at the end of the whole post-processing. Strangely, when the final bake was not achieved, the drift in the nMOS transistors appeared unchanged but the drift in the pMOS was increased up to +25 mV.

Discussion

Threshold voltage shift to the left measured after the evaporation and the following bake (step 6 in Table 4.1) strongly suggests that positive charges are trapped in the gate ox-

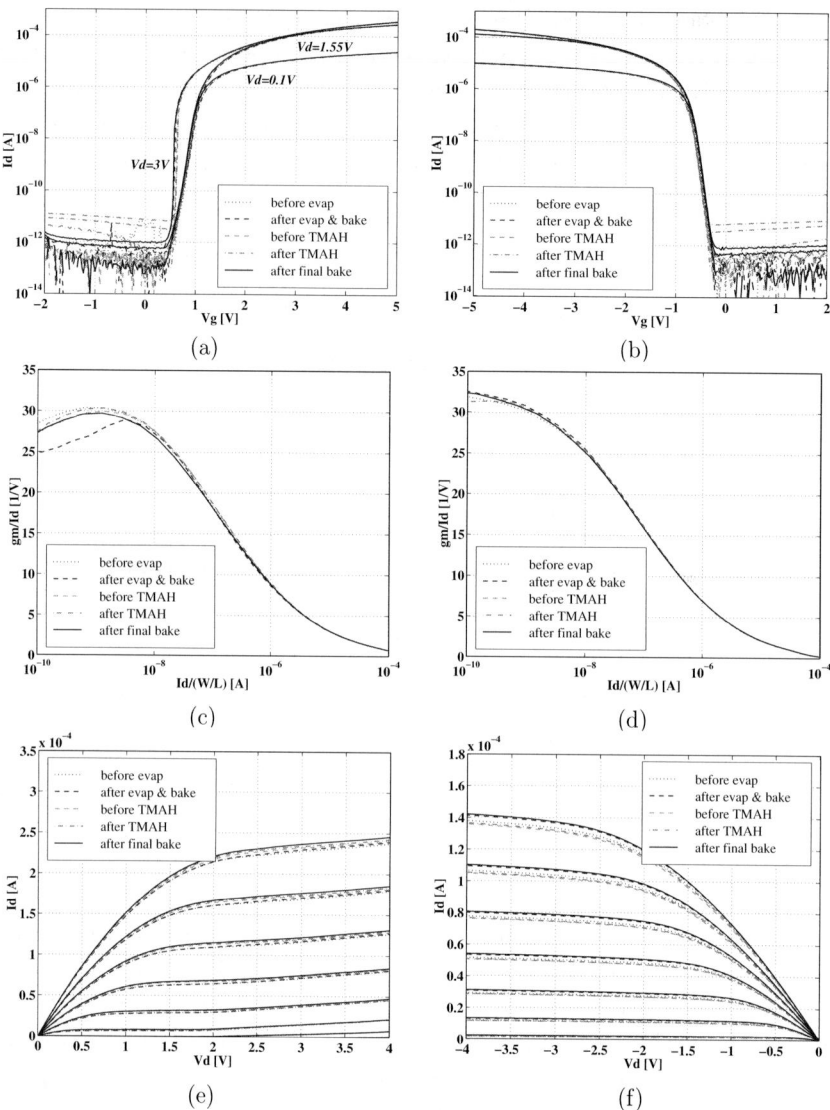

Figure 4.9: Measured characteristics during the post-processing steps for typical 4x3 μm^2 transistors. The nMOS with doped channel P2 are shown to the left while the pMOS with doped channel P1 are depicted to the right. The back gate was always fixed at 0 V. $I_d - V_g$ for (a) nMOS with V_d=0.1; 1.55 and 3 V and (b) pMOS with V_d=-0.1; -1.55 and -3 V; $g_m/I_d - I_d/(W/L)$ for (c) nMOS with $V_d = 1.55$ V and (d) with $V_d = -1.55$ V; $I_d - V_d$ for (e) nMOS with V_g varying from 0 to 4 V by steps of 0.5 V and (d) pMOS with V_g varying from 0 to -4 V by steps of -0.5 V.

4.4. MEASUREMENTS

ide during this stage. This assumption is confirmed in literature. E-beam evaporation is known to induce X-ray damage and even some ion damage on devices [27][178][184][185][182]. In such systems, a high-intensity electron beam gun is focused on the target material to allow its heating at temperature high enough for evaporation (see Fig. 4.10). X-rays can be generated due to the highly excited electrons in the material being evaporated decaying back to core levels [185].

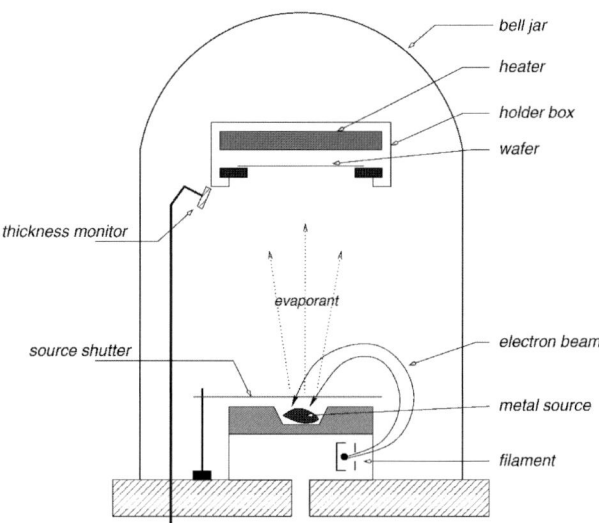

Figure 4.10: Typical electron-beam evaporation set-up like the VACOTEC used here (from [27] and VACOTEC).

These high energy particles induced by the evaporation can generate high densities of electron-holes pairs in an oxide [186][187][188]. For a thermally grown gate oxide with an applied positive field, electrons that escape initial recombination will rapidly drift toward the gate and holes will drift toward the Si/SiO$_2$ interface. As unrecombined holes approach the Si/SiO$_2$ interface, some fraction of the holes will be trapped, forming a net positive oxide trapped charge Q_{ot} for both pMOS and nMOS transistors. The large concentration of oxide trapped charges results not only in threshold voltage shifts (to the left) but also in sneak paths for current flow which lead to higher leakage currents. Interface traps Q_{it} also generated at the Si/SiO$_2$ interface can reduce the carrier mobility and cause a shift of the threshold voltage. Interface traps are usually negative in nMOS transistors and positive in pMOS [187][183]. Combined with the positive charge due to hole trapping, interface traps offset the hole trapping charge for nMOS devices, and the two effects add for pMOS devices [187].

Quick measurements just after the Ti-Pt evaporation (see Table 4.3 for deposition conditions) -before the first annealing- revealed strong threshold voltage drifts as well as very high leakage current (higher than 10^{-6} A). In the worst cases, the transistors were always passing. Same observations were reported in [189] after exposure of transistors (nMOS 20x0.6 μm^2) to radiations of about 10 krad. The same kinds of measurements were performed on an other wafer after Gold evaporation. Same observations were done while not so significant. Gold is indeed evaporated at a much lower temperature than Ti and Pt. Therefore less radiations are generated during evaporation, leading to less damages in the transistors characteristics.

Table 4.3: Pt and Ti evaporation conditions.

Type	Current	Extraction voltage	Deposition rate	Deposition vacuum
	[mA]	[kV]	[Å/sec]	[mbar]
Ti	54	8	1	1.4×10^{-6}
Pt	125	8	1	1.4×10^{-6}

As suggested in [184] and [182], annealing under Forming Gas at 432°C during 30 minutes was performed and removed the major part of the trapped charges in the gate oxide and therefore returned the threshold voltage and the leakage currents almost to normal. Nevertheless, we noted some limitations of this method, especially in large transistors as we will discuss further. This process was performed without special care on wafers covered with Ti-Pt interdigitated electrodes thanks to their high CMOS compatibility level. On the contrary, the wafers covered with Gold were annealed up to 320°C only to avoid diffusion problems of gold into the silicon, during 30 minutes under nitrogen outside the clean-room to avoid contaminations problems. The recovering was not so successful in this case.

To explain the difference of threshold voltage recoverings between the 20x20 and the 3x3 μm^2 transistors after the first annealing, we can not put forward the difference of charges density in the oxide due to the difference of area. Indeed, the trapped oxide charges are always reported to the surface area and are therefore included without dimensions into the calculation of the threshold voltage. A more probable assumption would be that hydrogen coming from the forming gas could not diffuse in the whole oxide volume of

4.4. MEASUREMENTS

the 20x20 μm^2 transistors to react chemically with traps and render them electrically inactive. The large polysilicon gate may avoid the hydrogen to diffuse such as the nitride as mentioned in [181]. On the contrary, hydrogen can diffuse up to the center of the channel in 3x3 μm^2 transistors. Perhaps a longer annealing could allow a better diffusion in 20x20 μm^2 transistors. It was partially confirmed by the second annealing performed at the end of the post-processing.

An interesting comment can be done from the observations reported in [187]. Annealing would suppress the trapped holes in oxides but the interface traps would remain unchanged. This implies that the smaller effect of negative charges (in nMOS) from the interface traps begins to dominate over time while the positive trapped charges lessen. This behaviour can favorably readjust the threshold voltages and can partially explain the recovering observed in our case. In pMOS, the trapped positive charges decrease while the amount of positive interface traps remain constant. Therefore, the annealing contributes also to reduce the threshold voltage drift. This theory could explain the smaller threshold voltage drift observed in the 20x20 μm^2 nMOS transistors than in the pMOS ones after the first bake (see Fig. 4.4(a)). It also validates the smaller mobility decrease observed in the nMOS transistors than in the pMOS as observed after the first bake on the g_m/I_d curves of the 20x20 μm^2 transistors (see Fig. 4.7).

In SOI technology, there is a strong probability that charges were also trapped into the buried oxide. It is reported in [186] and [190] that trapped charges in the buried oxide produce additional effects in the same direction than the front oxide charges, on the electrical characteristics of the transistors. Nevertheless, a distinction is reported between partially and fully depleted devices. In the first ones, there is no interaction between the top and the back while the coupling is higher in the second ones. In this last case, the top gate transistor is electrically coupled to the back gate transistor and charge trapping in the buried oxide directly affects the top gate transistor characteristics.

This behaviour was confirmed in our case. An enhanced degradation of the threshold voltage was observed in fully depleted transistors. In 20x20 μm^2 transistors, the partial recovering of the threshold voltage after the first annealing (step 6 in Table 4.1) confirms that trapped charges stay despite of the annealing. It seems not the case in 3x3 μm^2 transistors. As observed in Fig. 4.4(a), the threshold voltage drifts appeared sligthly decreasing when we increased the doping level in nMOS transistors. When the channel was undoped (intrinsic) the nMOS is fully depleted and showed a maximum voltage shift. The shift decreased at doping P1 and remains constant from doping P2. At these two last doping levels, the nMOS transistor is nearly fully depleted and shows a lower coupling between the back and the top gate. On the contrary, our previous measurements confirmed that the pMOS transistors are fully depleted. In their case, the threshold voltage drift increased when increasing the doping level. The positive trapped charges in the oxides or at the interfaces contributed to increase the holes conduction.

The other post-processing steps achieved after the evaporation did not significantly change any more the threshold voltage or other characteristics. Nevertheless, we observed a slight increase of the leakage current after the TMAH etching in transistors of both sizes. This result strongly suggests that the silicon etchant introduced some defects from the back of the wafer. This assumption will be confirmed further with the gated diodes measurements. The final bake offsets this increase in the smaller transistors (Fig. 4.6) but increases more the leakage current in larger transistors (Fig. 4.5). The final bake probably rendered these defects electrically inactive in smaller transistors but was not efficient enough in larger ones due to the bad penetration of hydrogen atoms to the center of the channel. One explanation for this leakage current increase after the final bake of the 20x20 μm^2 transistors could be the following: the high temperature of the final annealing allows the defects, which were not deactivated by hydrogen during the first annealing, to migrate in the oxide and to react with the channel/buried-SiO_2 interface. Consequently, the conduction was increased on the back and leakage current increased.

In the same way, we also observed strange variations in the g_m/I_d curves of the 20x20 μm^2 pMOS transistor working in strong inversion after the final bake. The g_m/I_d value remained constant after the TMAH etching but increased strongly after the final bake (Fig. 4.7(b)). The same assumption could be done. The trapped charges not deactivated during the final bake migrated to the interfaces. These positive charges increased the holes conduction in the channel, their mobility and therefore the transconductance over drain current. This could also explain the very small threshold voltage recovering observed after the final bake of the pMOS transistors (Fig. 4.5(b)) in comparison with the nMOS ones. This behaviour was observed neither in 20x20 μm^2 nMOS transistors nor in 3x3 μm^2 nMOS and pMOS transistors.

4.4.2 MOS capacitors

Capacitors were used to extract the MOS capacitance between the polysilicon gate and the source or drain via the doped channel. nMOS capacitance was investigated between

4.4. MEASUREMENTS

gate and source (or drain) of N$^+$ type via the gate oxide and the channel doped P2 as well as the pMOS capacitance between gate and source (or drain) of P$^+$ type via the gate oxide and the channel doped P1. The capacitors were designed with a surface of 0.11 mm^2 and a perimeter of 47.3 mm^2. During measurements, 0 V was applied on the source (or drain) whereas the gate voltage was increased step by step.

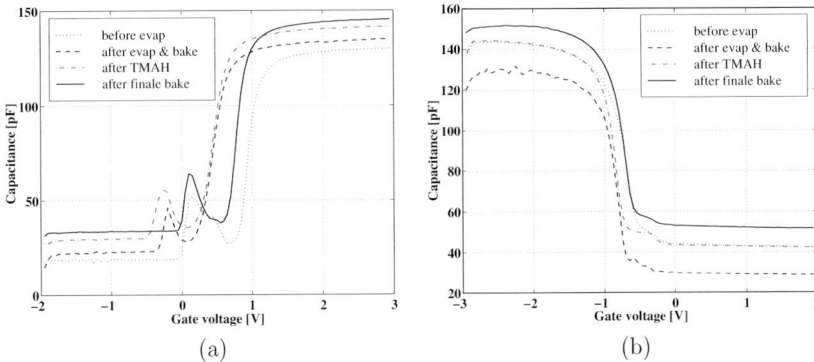

Figure 4.11: Measured $C - V$ characteristics for (a) nMOS capacitor between the polysilicon gate and the source (or drain) N$^+$ via the doped channel P2 and (b) pMOS capacitor between the polysilicon gate and the source (or drain) P$^+$ via the P1 doped channel. 0 V was applied on the source during measurements.

Capacitance measurements confirmed the threshold voltage drift to the left after the evaporation and the first annealing in both capacitors types. This drift was partially recovered after the final bake for the N type capacitors and fully recovered for the P type ones. Regarding the capacitance values, each post-processing step seems to increase more the capacitance. Nevertheless, this increasing could be due to the introduction of parasitic capacitances in parallel during measurements since the increasing appeared the same for negative and positive gate voltages. The same observation was done on other capacitances measurements. Measurements of the gate current also confirmed this increase.

The threshold voltage drift which appears after the evaporation and the first bake was previously explained. The capacitance coupling appears earlier due to the presence of positive charges in the gate oxide. On the contrary, in the pMOS capacitor, positive charges in the gate oxide contributes to a later formation of the channel. On the other hand, it is difficult to explain the different shift values we observe between both capacitors types as well as the complete recovering in the pMOS capacitor and the partial one in the nMOS. The fact that the channel seems partially depleted in first case and fully

232 CHAPTER 4. SOI-CMOS COMPATIBILITY VALIDATION

depleted in the second one can not explain this behaviour. We assumed above that large gates avert the hydrogen to diffuse in the whole gate oxide volume. Our capacitors were designed in fingers shape, each of them featuring width of 4 μm. The bake was therefore more efficient in this case than in the case of 20x20 transistors and can already explain the higher recovering achieved by the final bake. Nevertheless, the recovering difference remains unexplained at this time.

It can be observed that the $C-V$ characteristics are not complete. For the nMOS capacitor, the capacitance reaches its maximum in strong inversion, when the channel at the silicon/oxide interface is complete. On the contrary, the capacitor in accumulation revealed a very low value due to the lack of carriers and then, of an accumulation layer. The bump observed before the capacitance increase to the inversion domain seems not consistent. It appeared and disappeared from one to another measurements and is influenced by the measurement from the positive values to the negative values on the gate. A body contact in this case could supply enough holes to allow the formation of a complete accumulation layer [173]. In this case, the capacitance in accumulation would be increased, the $C-V$ characteristic would be complete and the transient floating body effect would be suppressed.

4.4.3 Gated Diodes

Gated diodes with two independent front and back gates can be advantageously used to extract the recombination parameters such as the carrier recombination lifetime and surface recombination velocity in SOI CMOS devices. Using the conventional front-gate and controlling the bias of the substrate, it is possible to decouple the recombination mechanisms in the volume of the Si film and at the front and back Si-SiO$_2$ interfaces [191].

Measurements of this kind of gated-diodes allowed us to study the impact of our post-processing steps on the quality of the silicon film and the interfaces in our devices.

All measurements were achieved under a small forward bias (0.2 V) as suggested in [191]. The devices we used were NIP gated diodes in meander shape with a gate size W/L of 3/852 μm. The forward current was measured as a function of the front-gate voltage at various back-gate voltages (from –15 to 15 V by step of 5 V) and as a function of the back-gate voltage at various front-gate voltages (from –3 to 3 V by step of 1 V). Experimental results are shown in Fig. 4.12(a) and (b) after the 4 post-process steps.

4.4. MEASUREMENTS

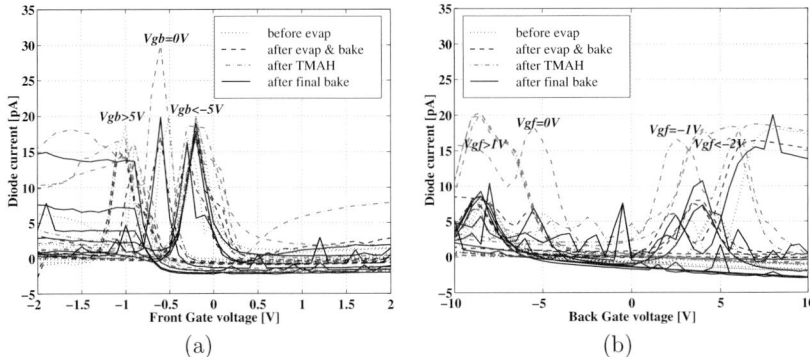

Figure 4.12: Measured forward current in gated-diodes ($W/L=3/852$) versus (a) the front gate voltage for various back-gate voltage (step of 5 V), and (b) the back-gate voltage with a front gate voltage step equal 1 V. The forward bias was fixed at 0.2 V.

From our experimental results, simulations and the detailed study of [191], we extracted the following observations. All curves showed the same behaviour at any post-processing step excepted after the TMAH etching. According to the works of [191], the largest peak in Fig. 4.12(a), which is observed when both Si film surfaces are depleted, represents the overall recombination current involving recombination at the front- and back Si-SiO$_2$ interfaces and in the Si film volume. The higher peak after TMAH etching reveals that one of these components was changed. When the back-gate is kept in accumulation ($V_{gb} < -5V$) or inversion ($V_{gb} > 5V$), the peaks result from the front-surface recombination. It appears unchanged in our case. Finally, when the front interface is in strong accumulation ($V_{gf} < -1V$) or inversion ($V_{gf} > 0.5V$), a plateau appears to the left and to the right of the peaks and is due to the back-surface component. It appears higher in our case after TMAH etching revealing in first hypothesis a higher back-surface recombination component after TMAH etching than before. This result is confirmed by the second measurements. The 2 largest peaks in Fig. 4.12(b) which appear when the front-gate is maintained in accumulation ($V_{gf} < -1V$) or inversion ($V_{gf} > 1V$), reveals the back-surface component. These peaks appears again higher after TMAH etching.

In conclusion, the two times higher recombination current measured after TMAH etching strongly confirms our previous assumption that TMAH introduces some defects at the back interface which participate to the conduction. This observation is quite consistent since the etchant is in direct contact with the back Si substrate and could drive impurities through the substrate up to the buried oxide. Nevertheless, these impurities completely vanished (or were dispersed) after the final bake.

4.4.4 Summary

We put forward a number of assumptions in order to try explaining the main behaviours observed in our measured curves. Nevertheless, further investigations are required for a deeper understanding of all the reported phenomena. But it would necessitate a full book on the subject. For instance, other measurements could be done to separate the effects of interface from oxide trapped charge densities in transistors as reported in [192]. But in SOI, it is more complicated due to the two oxide layers.

It appeared from our measurements that finally the metallic interdigitated electrodes evaporation is the main cause of changes in the electrical characteristics of our tested devices. The evaporation introduces significant threshold voltage drifts and mobility decrease in larger transistors despite of their annealing under forming gas. In smaller ones (sizes of 3x3 or 4x3 μm^2), the annealing was efficient enough to return the devices to normal. The TMAH etching also introduces some defects on backside which were responsible of a leakage current increasing. Nevertheless, we demonstrated that once again, a final annealing suppresses all these defects in smaller transistors.

We demonstrated the whole SOI CMOS compatibility of our process for sufficiently small transistors (with width/length ratio equal to about 4/3). If larger transistors need to be integrated, it could be interesting to depose the metallic layer by sputtering way instead of evaporation. Unfortunately, the sputtering of platinum is more expensive than its evaporation. Furthermore, it is reported that it generates UV-rays which could also damage the devices [179]. Nevertheless, this deposition technique is more common in post-process of sensors [193] and could be advantageously tried in the future.

In our process, our devices were not covered by any passivation layer which may protect them from radiations. We previously explained that a pyrox oxide used as passivation layer could affect the residual stresses equilibrium in our membranes since it could not be etched selectively with the densified PECVD on the top of the membranes. The use of PECVD nitride instead of oxide would be attractive but [182] reported that PECVD nitride is not so efficient as barrier to the radiations. Furthermore, nitride blocks the diffusion of hydrogen during post-annealing and avoids the reduction of trapped charges occurring during the evaporation.

The best way to increase the radiation hardening of our devices would be to decrease their sizes, in width as well as in thicknesses. It is reported in [182] that the relation between threshold voltage shift and gate oxide thickness t_{ox} is a power law, $(t_{ox})^n$, n depending on the thickness and the oxide type (wet or dry). Ultra-thin gate oxides (15-30 nm) are less vulnerable to radiations [188]. Their higher hardness is possibly a result of trapped-hole annihilation by electrons tunneling from the oxide-silicon interface and the polysilicon gate [188].

4.5 Transistors on membrane as final demonstrator

nMOS and pMOS 4x3 μm^2 transistors were fabricated on membrane and tested as a final demonstration of the full CMOS compatibility of our process. Up to this point, we demonstrated the full compatibility of the post-processing according to the standard IC fabrication. In this section, we will show that it is possible to combine some intermediate processing steps during the standard processing with a complete post-processing without damaging the compatibility. The aim of this design is to be able in the future to integrate e.g. pressure sensors using SOI CMOS transistors on a thin dielectric membrane as sensing element. By replacing the polysilicon gate by a gas-sensitive film, our transistors on membrane could also be advantageously transformed into MOSFET Gas sensors (also called gasFET). These two applications of this design will be briefly described further. An other application we will not discuss here would consist in replacing transistors by diodes as thermometer in order to check more precisely the temperature of the heater. In case of flow sensors, thermopiles could be then replaced by accurate diodes to improve the flow sensitivity.

4.5.1 Design and fabrication

The transistors on membrane were fabricated following the same process as the one used to build the microheater on membrane in SOI technology (see chapter 1 of this part). The main difference is that we left a silicon island un-etched at the middle of the membrane in order to integrate the transistor. The following pictures (Fig. 4.13, 4.14 and 4.15) give some illustrations of the transistor in order to understand the design of the device.

It must be noted that the oblique membrane seen in Fig. 4.13 is due to a flat uncorrectly aligned to the crystallographic [110] direction (the whole process was aligned to the

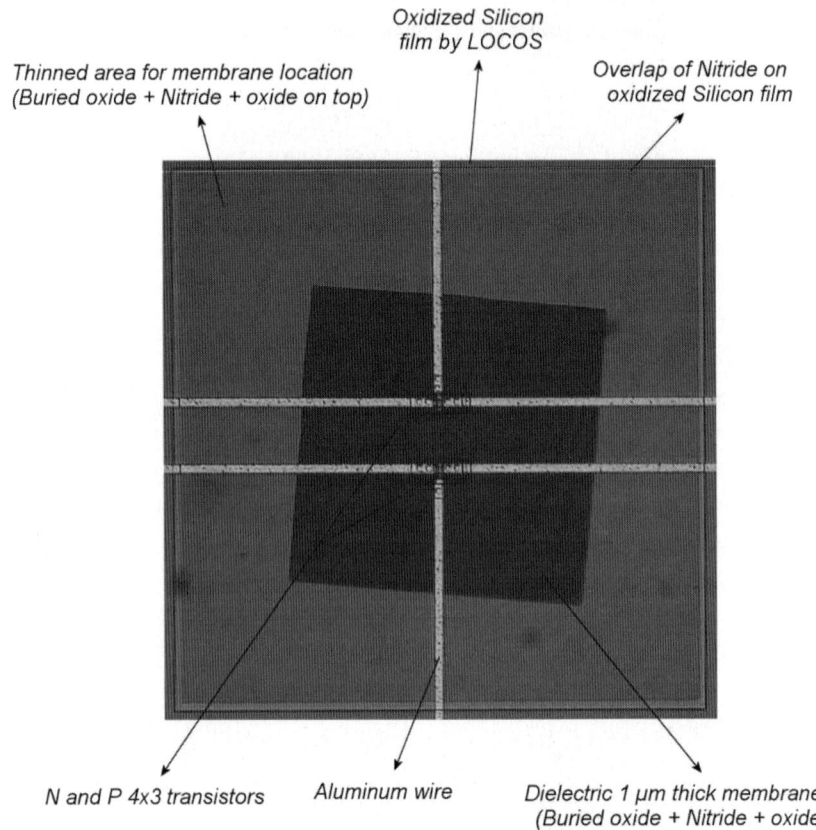

Figure 4.13: Picture of the whole thinned area surrounding the 1 μm thick dielectric membrane which supports n and pMOS 4x3 transistor. The released membrane shown here has a size of about 180x180 μm².

4.5. TRANSISTORS ON MEMBRANE AS FINAL DEMONSTRATOR

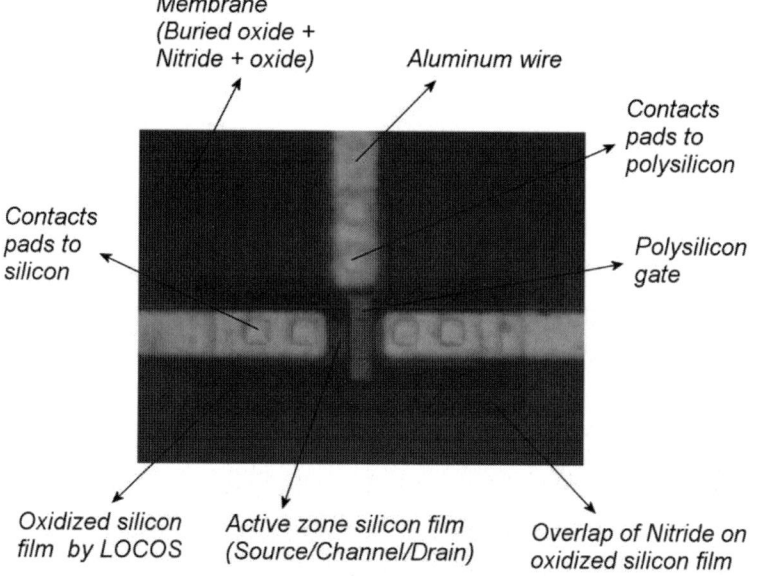

Figure 4.14: Close view of a transistor on a dielectric membrane in SOI technology.

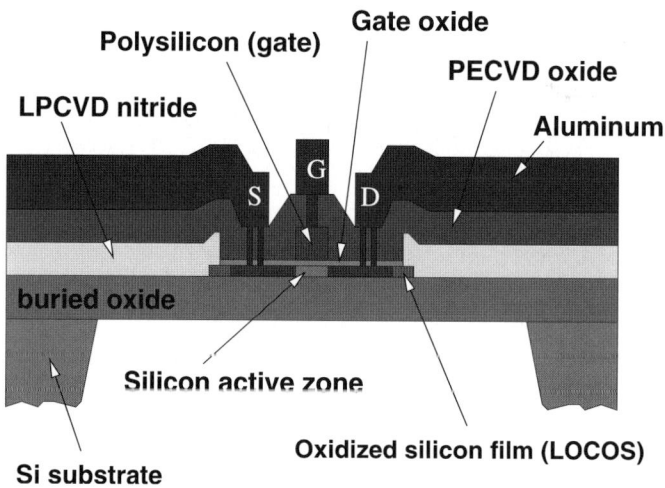

Figure 4.15: Cross section of a transistor on a dielectric membrane in SOI technology.

flat, but during the laser cut of these 3" wafers from 8" SOI substrates, the flat was not correctly sliced). The overlap of nitride can be seen at the boundary between the circuits part on the membrane and the thinned area for the membrane location. Much closer to one of the two transistors (Fig. 4.14), we can observe an other overlap of the nitride layer on the oxidized silicon film (LOCOS) in order to avoid any uncovered trench in case of bad alignment. This is illustrated in the cross sectional view in Fig. 4.15.

4.5.2 Measurements and discussion

Measurements of the NP2N and PP1P transistors on membrane revealed their excellent characteristics. Measurements were performed on several identical devices distributed over the same wafer before and after the TMAH etching. We had also the possibility to compare directly on the same wafer, identically-designed transistors on the substrate and on membranes.

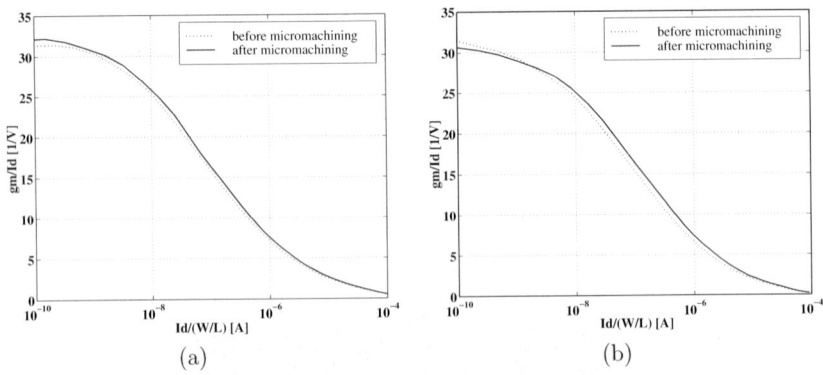

Figure 4.16: $gm/I_d - I_d/(W/L)$ measured characteristics before and after TMAH etching for (a) nMOS transistors with $V_d = 1.55$ V and (b) pMOS transistors with $V_d = -1.55$ V. The back gate was fixed at 0 V.

First of all, a better mobility was provided as we extracted from the gm/I_d and $I_d - V_d$ curves (Fig. 4.16 and 4.20). Same observations were done in Silicon-on-nothing (SON) devices [194].

Surprisingly, the back gate voltage had the same impact on the $I_d - V_g$ curves with and without silicon on backside. It is not possible that a part of silicon is left over on backside of the membrane to conduct the current. Fig. 4.18 shows the nMOS and pMOS transistors by transparency from the backside.

4.5. TRANSISTORS ON MEMBRANE AS FINAL DEMONSTRATOR

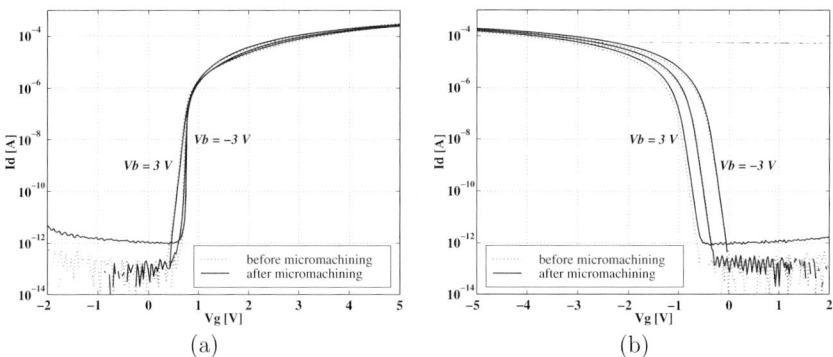

Figure 4.17: $I_d - V_g$ measured characteristics before and after TMAH etching for (a) nMOS transistors with V_d=3 V and (b) pMOS transistors with V_d=-3 V with back gate voltage = -3; 0 and 3 V.

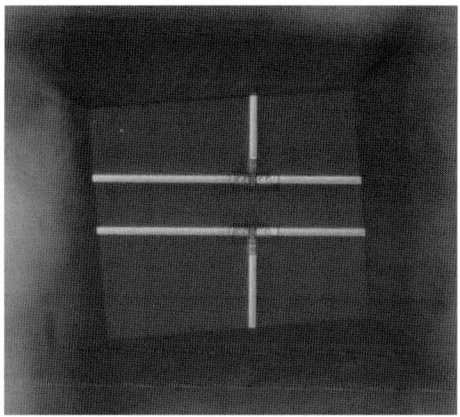

Figure 4.18: Picture from the backside of the membrane showing the two transistors (nMOS and pMOS) by transparency through the 1 μm thick dielectric membrane.

Nevertheless, a capacitance coupling appears between the chuck supporting the wafer during measurements and the buried oxide which could fix the voltage on the backside of the buried oxide. Electrical field lines between the silicon edges of the membrane could also fix the bias on backside. So, a capacitance in parallel with a conductance could connect the silicon as well as the chuck to the membrane.

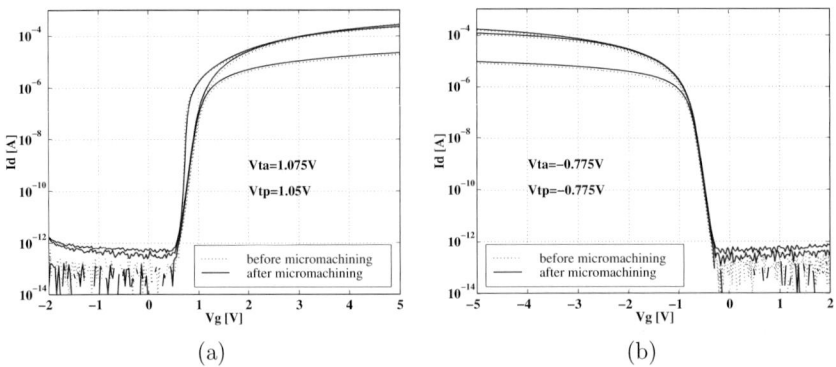

Figure 4.19: $I_d - V_g$ measured characteristics before and after TMAH etching for (a) nMOS transistors with V_d=0.1; 1.55 and 3 V and (b) pMOS transistors with V_d=-0.1; -1.55 and -3 V. The back gate was fixed at 0 V.

The leakage current measured on 4x3 transistors after TMAH etching appeared lower when silicon was removed underneath the transistor than when silicon remained not etched on backside. It could confirm the appearance of some defects at the substrate/buried oxide interface during TMAH etching which were eliminated when silicon was removed on backside.

Regarding the $I_d - V_d$ curves, an interesting observation can be done. For gate voltage values higher than 3.5 V, the kink effect vanishes. Some hypotheses can be done to explain this behaviour but the main probable effect is the self-heating effect appearing at high drain and gate voltages. The thermally insulated device on the membrane can reach higher temperatures at a given voltage than if it is on substrate which therefore leads to a strong drain current decrease by self-heating. This observation was confirmed in [130] where SOI transistors insulated from the substrate followed exactly the same behaviour in the similar case. A temperature rise decreases the carrier mobility and therefore the drain current.

4.5. TRANSISTORS ON MEMBRANE AS FINAL DEMONSTRATOR

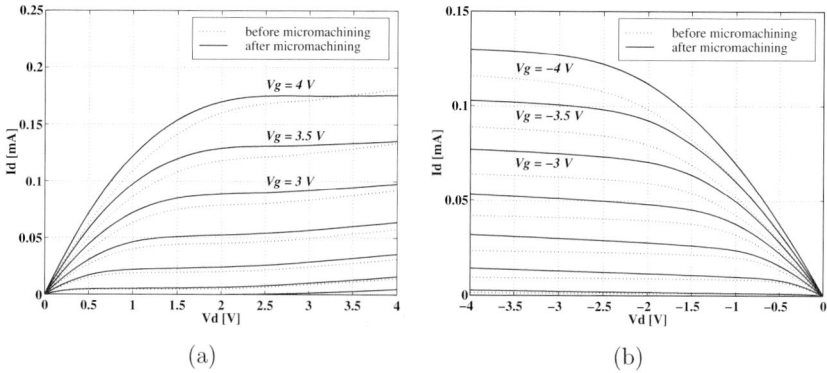

Figure 4.20: $I_d - V_d$ measured characteristics before and after TMAH etching for (a) nMOS transistors with V_g varying from 0 to 4 V by steps of 0.5 V and (b) pMOS transistors with V_g varying from 0 to -4 V by steps of -0.5 V. The back gate was fixed at 0 V.

4.5.3 Applications

Pressure sensor

The first purpose of this design is its use as pressure sensor. The use of MOS transistor with the channel placed on a membrane, near to the edge, longitudinal, and transversal (perpendicular) to the edge, allows to sense pressure [18]. The membrane deflection leads to surface stress in the channel and thereby modulates the carrier mobility due to piezoresistive effect [195]. Consequently, the variation of the source-drain current of the MOS transistor becomes pressure dependent. The piezoresistive coefficients of a silicon inversion layer are reported as high as those of diffused piezoresistors [196] but [18] reports a better piezoresistive sensibility to pressure with P channel transistors and (100) orientation substrates.

The principal disadvantages of the piezoresitive sensors are their high thermal instability, self-heating and an abrupt decrease in the device sensibility due to the misalignement of the resistors with respect to the diaphragm edges. This type of sensors therefore require thermistors in the readout circuit to compensate for temperature variations and laser trimming to precisely balance the resistive bridge [18]. In comparison with classical polysilicon or diffused piezoresistors, transistors can be made very small and precisely positioned where the stress distribution is maximum. The ability of combining several MOS transistors is also very attractive: a configuration similar to a Wheatstone bridge can be realized, increasing the detection sensitivity while reducing temperature and noise effects. Finally, transistors can be mounted in an inverter configuration in order to enable a direct conversion of the mobility variation into an output voltage [196].

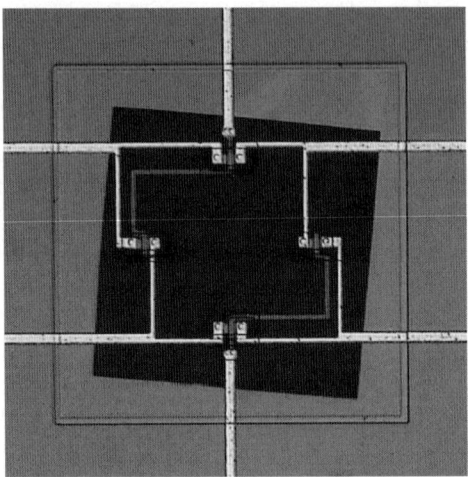

Figure 4.21: Pressure sensor with transversal and longitudinal pMOS transistors as transducers on dielectric membrane in SOI technology.

A typical prototype (as shown in Fig. 4.21) was designed and fabricated with pMOS transistors following the same processing as the one explained above. Unfortunately, the misalignement of the flat with the [110] direction as well as the membrane size larger than expected (the channel is on the membrane instead of being on its border) need some improvements in a future process to make our sensor fully effective.

Gas sensor

A gasFET sensor could also be fabricated from this design. In this kind of sensors, the gate oxide insulator has to remain uncovered at the end of the transistor processing. A gas-sensitive film, such as a catalytic metal, a metal-oxide or a polymer, is then deposited on top of the insulator to form the gate of the transistor [197]. The gasFET is based on the change in the threshold voltage measured on the gas-sensitive gate in the presence of a gas [3]. The catalytic gates require temperatures of at least 180°C-200°C to allow the reaction to occur and therefore, to reduce the power consumption, the sensor needs to be suspended on a silicon membrane [3]. Usually, the MOSFET gas sensor is built into a silicon island suspended by a dielectric membrane as reported in [197] and [198]. The membrane is in this case heated to a target temperature by a semiconducting resistor surrounding the MOSFET area. Temperatures of 170°C are reached with a power consumption of 90 mW [197]. As suggested in [197] and [198], SOI technology is a better candidate to increase the working temperature of these sensors and to simplify the silicon island processing.

[3] and [130] reported a resistive gas sensor not heated by a classical resistor but by a MOS transistor, in SOI technology. In this case, the MOS microheater is integrated in the SOI silicon film, the meandered polysilicon gate playing the role of hotplate. The membrane is constituted by the stacking of the buried oxide, the LOCOS oxide outside the active area and a passivation layer. Intedigitated electrodes as well as an sensing layer are added on top. Temperatures up to 200°C are reached by this way with a DC power consumption of *ca.* 75 mW [130].

We could combine the gasFET design, the vertical design of the SOI MOS microhotplate and our polysilicon loop shape microhotplate with our SOI transistors on membrane to build a new successful gasFET sensor in SOI technology. In this case, we would integrate the gasFET transistor in the middle of our polysilicon loop shape microhotplate and or new gas sensor would feature very low power consumption, high temperature as well as high sensitivity to gases.

4.6 Conclusions

We revealed in the introduction that the additive steps at the beginning of the standard CMOS SOI process (to provide the second layer of our membrane in LPCVD nitride) did not affect the performances of our transistors. This was expected since these steps were achieved far enough from the integrated circuits. With the transistors on membrane, we demonstrated that these processing steps can also be advantageously combined to the IC fabrication without altering the CMOS compatibility.

We demonstrated also the full compatibility of our post-processing not only in the case of gas flow sensors on membrane working close to the electronics but in the case of transistors suspended on membrane too. Nevertheless, we showed and discussed the limitations of the full compatibility when metallic layers need to be evaporated near to large electronics. Our measurements revealed also that TMAH introduces a slight leakage current increase which can be canceled by a dedicated annealing.

Therefore, the full SOI CMOS compatibility of our process was demonstrated. This confirms that our gas flow sensors are ready to be co-integrated in the same chip to feature smart microsystems. Furthermore, we introduced for the first time the ability to co-integrate high performance transistors on dielectric membrane, which can open the door to a large kind of sensors applications, such as pressure, gasFET or temperature sensors.

Part IV

Conclusions and outlook

This work has tackled the **co-integration**, on the same chip, of sensors on thin film membranes and CMOS circuits needed for their signal processing. We focused our design and fabrication on the most advanced of the CMOS processes, namely **Silicon-on-Insulator (SOI)** technology. In the introductive part, we explained how co-integration can be advantageously achieved with this technology in comparison with other hybrid techniques. We showed that despite the higher cost of the starting SOI material and the additional processing steps required by the co-integration, Silicon-on-Insulator technology can provide high performance co-integrated sensors without increasing the design efforts. In addition, packaging problems are reduced and co-integration increases the long-term reliability as well as miniaturization, which in the end results in a higher quality system. Furthermore, the high temperature and radiation hardness of SOI technology make it the best option to achieve co-integrated microsystems dedicated to work in harsh environment, such as automotive, aerospace and industrial sectors. The overall cost of co-integration is thereby far from being always increased in comparison to hybrid techniques. In addition, we show also that if in some cases the co-integration can be an option, it is imperative in other applications. We especially report the main three applications of a thin co-integrated dielectric membrane: thermal insulation, electrical insulation and support for mechanical sensing.

The content of our work has been divided in two main parts, the first describing the materials and techniques developed to optimize our sensors performances as well as to allow their co-integration, the second being centered on the sensors themselves and the validation of their full CMOS compatibility.

In the part II of this work, a thin **dielectric membrane** has been built step by step. After an original review on the **TMAH etching** and a large variation of tests to illustrate its main interesting properties, this micromachining technique has been optimized to make it the best etchant adapted to release dielectric membranes in post-processing of a SOI-CMOS standard process. In particular, its etching performances have been optimized in comparison with the ones reported in the literature while ensuring a great selectivity towards aluminum. The selectivity towards other metals has been also investigated for the first time and has been revealed critical when interdigitated sensing electrodes are to be placed on top of our membrane. Its CMOS compatibility has been fully confirmed by electrical measurements on various Si devices integrated close to the micromachined membranes. Gated diodes particularly revealed for the first time the presence of some positive charges trapped into the buried oxide or at the Si-film/buried-oxide interface as a result of TMAH etching. Fortunately, these traps appeared fully suppressed after a dedicated post-annealing.

In addition, the ability of the TMAH anisotropic etchant to under-etch masking materials featuring specific geometries has been explained and allowed us to micromachine suspended reliable structures (metallic or dielectric) such as micro-bridges, micro-rings and cantilevers. Such mechanical microstructures -well known to extract the **residual stress** in polysilicon or nitride films- have been used for the first time to measure the residual stress in oxide and aluminum films by using the silicon as sacrificial layer. Residual stress in thermal oxide, LPCVD nitride and PECVD oxide films has been measured by these means and compared with the values obtained with the well-established wafer curvate method. Results matched and both methods have been shown complementary. Furthermore, substrate curvature method has been expanded to the measurement of the overall residual stress in a stack of layers and revealed that slightly tensile stacking can be targeted. Finally cantilevers allowed us to extract the stress gradient in each separate layer as well as the strain gradient in their stacking. Since the residual stress is strongly responsible for the planarity and the robustness of a membrane, our extracted results allowed us to design a robust and flat stacked dielectric membrane (only 1 μm thick) starting from the buried oxide layer inherent in the SOI technology. Target values have been confirmed after a complete gas sensor processing and allowed us to analyze the impact of each processing steps on the overall average stress of our membrane. The combination of both techniques thereby appeared as a new original way to design robust membranes with well-controlled residual stresses as well as to extract locally the stress during processing.

The part III of our work has been based on our optimized membranes (in shape and robustness) and on the technique used to release them (without damaging the surrounding circuits) for building microsensors. **Microheater-based sensors**, especially gas and flow sensors, have been designed and fabricated. We have shown that despite of the considerable amount of works already performed in this field, our study proposes original designs in SOI technology; their CMOS compatibility has been validated and we report very low power and low cost results thanks to the use of materials inherent to the integrated circuits processes. Our careful review has revealed that most of the sensor fabrications referred as "CMOS compatible" have been in fact rarely produced in standard CMOS processes with integrated circuits, or used too complicated process steps (increasing the fabrication cost). Therefore, we choosed to use gate polysilicon as microheater material despite of its well know drawbacks regarding thermal stability over time at high temperatures (from 400°C). Thanks to its fair temperature coefficient of resistance, polysilicon resistor has also been used as thermometer after appropriate calibrations. A model has been developed in order to avoid tedious separate calibrations of each microhotplate featuring the same design.

Our microhotplate-based gas sensor has firstly been simulated in order to target high uniformity, high thermal insulation and low power consumption when heating gas sensitive layers at temperature up to 400°C. A novel polysilicon loop shape has been designed as well as a membrane as thin as 1 μm to reach high performances. Various electrical measurements allowed us to extract the power consumption for different membrane sizes and loop geometries and revealed that our microhotplate featured the lowest power consumption we can find in literature thanks to its thin dielectric membrane and optimized design. Unfortunately, the very low power consumption has been counteracted by a too high bias needed to reach the operating temperature (20 V at 400°C) which complicates the integration of its control electronics. An improved design which consists in connecting 2 microheaters in parallel has been discussed in order to decrease this bias to a more appropriate value. Furthermore, measurements of microhotplates surrounded by thermopiles and polysilicon resistors (such as in flow sensors) revealed additional interesting informations about the thermal gradient throughout the membrane as well as the response time of the microheater. In addition, thermal uniformity has been controlled after fabrication by an original dedicated thermoreflectometry technique developed for this purpose and based on additional polysilicon pads uniformly distributed within the membrane to increase the measurement resolution. Unfortunately, these measurements revealed unsatisfactory thermal uniformity due to cold points on the accesses. Perpendicularly to the accesses, variations up to 50°C have been measured while higher ones have been extracted from the accesses to their opposite. An improved design has been discussed for a future development.

In order to expand our microheater towards a full **gas sensor**, active films as well as interdigitated electrodes are needed to sense gases by extracting the gas dependent conductance variations of the sensitive film. Furthermore, we outlined the fact that gas sensitive layers need to be heated up to 700°C to guarantee a greater stability at their operating temperature. The metal choice for the sensing electrodes has been discussed and our tests revealed that the evaporation of platinum on titanium constituted a good compromise to feature low cost, high operating temperature, fair CMOS compatibility, good adhesion and process easiness. Gold has been rejected due to its too low melting point and its weak CMOS compatibility. Transistors measurements directly performed after the metal evaporation revealed slight threshold voltage drifts and increasing leakage currents. We reported that these drifts are caused by oxide trapped charges due to X-rays induced by the e-beam evaporation. Annealing under forming gas revealed to be efficient to return the threshold voltage as well as the leakage current to normal. This has indeed been verified for small transistors. Unfortunately, larger devices are not so well recovered. A physical explanation of these observations has been discussed. We concluded that in order to decrease the radiation damages, sputtering or dedicated passivation layer could allow to prevent damages in large transistors despite of their respectively higher cost and higher process complexity.

Measurements at high temperature as well as during a long time (ageing) have been performed in order to understand the behaviour of the polysilicon film under these conditions. Some works in this field have been found in the literature but suffer from several deficiencies we tried to fill. We contribute to summarize these works and to extend them with new interesting data. We show that in situ heating of the gas sensitive layers at temperature as high as 700°C can be performed if it has been followed by pre-ageing tests during some days. By this way, the properties of the polysilicon returned to the ones it had before the high temperature heating and have been stabilized for future cyclings at the operating temperature. In order to avoid the tedious time-consuming pre-ageing tests, we suggested another solution for stabilizing the temperature which consists in monitoring the injected power during heating instead of only current or voltage.

Finally, three kinds of gas sensitive layers are investigated and our measurements in presence of various gases showed promising results and excellent selectivity to ethanol, ammonia and NO_2. Measurements in absence of gas have also been performed for reliability and ageing tests and confirmed the high thermal and mechanical robustness of our membrane.

The same microhotplate has been integrated between two polysilicon/aluminum ther-

Conclusions and outlook 251

mopiles in order to build a low cost CMOS compatible **flow sensor**. Our review of the recent published flow sensors results revealed once again that a lot of designs termed "CMOS compatible" have been rarely verified in CMOS processes. Microhotplate-based flow sensor has been compared with a simpler design based on a simple polysilicon wire between two thermopiles. Flow measurements confirmed the high performances of our devices, featuring low cost, fully CMOS compatibility, fair sensitivity on a large airflow rate range at a very low consumption and response time in the average of the published results. Unfortunately, the sensitivity of our thermopiles appeared very small as a result of low performance thermopiles but we showed that these could be easily increased in a process allowing two different polysilicon doping levels.

The last section of the microsensor part validates the **SOI-CMOS compatibility** of our previous processes. We already outlined the informations these measurements revealed after TMAH etching and interdigitated electrodes evaporation. Our measurements on co-integrated MOS transistors of different types and sizes, capacitors and gated diodes, after each critical step of our post-process, confirmed the full SOI-CMOS compatibility of our post-processing. Furthermore, the discussions of our measurements results constituted an original contribution. Indeed, the literature never reported so deeply the interactions between CMOS device characteristics and post-processing. In addition, we also showed that the additional steps of intermediate processing (patterning of the membrane location and nitride etching) needed to ensure the SOI-CMOS compatibility of our gas sensor has been neither time-consuming nor hazardous for the co-integration.

Finally, a novel powerful demonstrator consisting in n- and pMOS transistors insulated in a small silicon island on a thin dielectric membrane has been studied. This novel design probably constituted one of the most interesting contributions of this work. Not only because it does demonstrate in one device the full SOI-CMOS compatibility of our intermediate- and post-processes but it also opens the door to really innovative developments, such as MOS pressure sensors or thermodiodes-based high-sensitivity flow sensors. Furthermore, this design calls on all the advantages of the SOI technology, the co-integration and the dielectric membrane reported in the introduction. We summarize them as follows:

- easy processing thanks to the availability of the buried oxide as etch stop layer and the silicon film (patterned in silicon island shape) to integrate advanced transistors;

- electrical insulation and high frequency performances thanks to the dielectric membrane;

- mechanical sensing for pressure sensor;

- thermal insulation by the observation of self-heating effect on membrane;

- co-integration by the combination of CMOS circuitry with a typical sensor-fabrication technology, i.e. bulk micromachining;

- harsh environment applications conceivable for the complete co-integrated sensor.

Coming at the end, we hope that this small contribution to the large amount of publications in the field of sensors and circuits co-integration will improve the arguments to put forward the SOI technology as the big promise for the next generation of microsystems...

Appendices

Appendix A

(100) Silicon crystallography

From [27] [199] [11] [126] and [200].

The periodic arrangement of atoms in a crystal is called the lattice. All crystals are based on the repetition of a characteristic unit cell in a lattice, representative of the entire lattice. Semiconductor crystals are members of the cubic crystal family and have a unit cell based on the cube. To identify a plane or a direction in a lattice, a set of integers h, k and l called the *Miller indices* are used.

Considering an arbitrary plane in three-dimensional space having following equation,

$$hx + ky + lz = 1 \qquad (A.1)$$

Here, x, y, and z are any points in the plane and h, k, and l are the Miller indices which can be defined as the reciprocals of the intercepts formed by the plane with respectively x, y, and z axes. The numbers h, k, and l identify the plane uniquely, so the plane can be identified by the shorthand notation (hkl). Normal to the plane is a direction, whose projections onto the axes are h, k, and l. This direction is written $[hkl]$. Because a crystal is a repeating array, there will in fact be many planes parallel to our chosen one, with the same normal direction. In an ideal crystal, these planes are physically indistinguishable, and we can choose to represent all of them by reference to one plane whose Miller indices are a convenient combination of integers.

For a cube centered at the origin, the cube walls correspond to planes with intercepts such as $(1, \infty, \infty)$ or $(\infty, 1, \infty)$ or $(\infty, \infty, 1)$ resulting in indices (100) or (010) or (001). By taking the reciprocal of the intercepts, infinities (∞) are avoided in the plane identification. Planes (100), (010) and (001) as their corresponding directions [100], [010] and [001] are all crystallographically equivalent since they are symmetrical without regard to the distinction between x, y, and z axes. From any of these three directions, the cube is structurally (and therefore crystallographically) equivalent. This set of planes is denoted using braces, $\{100\}$, and their equivalent set of directions is identified with angle brackets, $\langle 100 \rangle$. Fig. A.1 depicts the most common orientations and corresponding planes: (100), (110) and (111) and their corresponding directions.

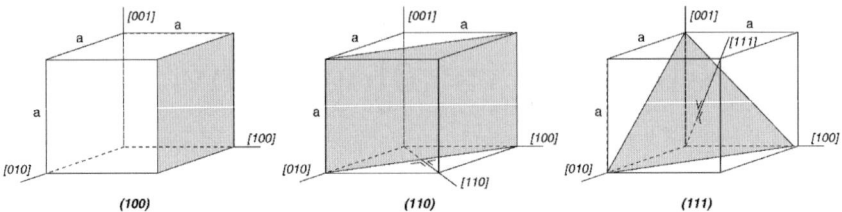

Figure A.1: Miller indices in a cubic lattice: planes and axes. Shaded planes are respectively (100), (110) and (111) (after [27]).

The simplest type of cubic lattice is the *simple cubic lattice.* This lattice is rarely found in nature. More common are the *face-centered cubic (FCC)* lattices having eight corners and six faces atoms. Common semiconductor crystals are based on an elaboration of the FCC lattice. In this structure, two face-centered lattices are interpenetrated, one displaced of $(0.25a, 0.25a, 0.25a)$ with respect to the other, were a is the length of one side of the cube (the value of a equals 5.4309 Å). When this lattice is built with identical atoms, the lattice is a *diamond lattice* as shown in Fig A.2. This is the crystal lattice of silicon.

In a diamond-cubic structure such as crystalline silicon, along the $\langle 100 \rangle$ directions, crystal planes are separated by length a (Fig. A.1); along the $\langle 110 \rangle$ directions by $0.707a$, and in the $\langle 111 \rangle$ directions by $0.597a$. Thus the $\{111\}$ planes present the highest packing density. It is interesting to finally note that a number of cavities of intersticial voids are located within the diamond lattice. Each of these voids is sufficient to contain an additional atom inserted when doping.

Silicon wafers are typically marked with flats, i.e. segments to indicate the orientation of the principal crystal planes. From one source to an other, a family of wafers can be identified by their orientations $\langle 100 \rangle$, of by their surface planes $\{100\}$. But more

APPENDIX A (100) SILICON CRYSTALLOGRAPHY

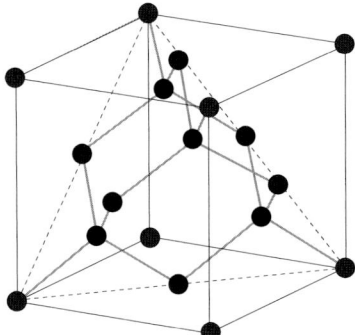

Figure A.2: Diamond-cubic structure of Silicon.

commonly, when we speak about one wafer, we consider its specific direction [100] perpendicular to its surface plane (100) or its surface plane itself. By the same way, the flat can be expressed as giving the direction [110] or being a part of the plane (110).

The most common orientations used in the IC industry are the ⟨100⟩ and ⟨111⟩ orientations; in micromachining, ⟨110⟩ wafers are also used quite often as well. But ⟨100⟩ wafers are adopted in most foundries as well as in our laboratory. We have considered therefore only silicon wafers with a ⟨100⟩ orientation. The flat of these wafers indicates the ⟨110⟩ direction. The precision on the flat is about 3°. A smaller secondary flat is generally used to identify the doping type of the wafer. In this way; on a p-type ⟨100⟩ oriented wafer, both flats are placed at 90°. Flat areas help orientations and doping determinations but also to the placement of wafers in fabrication equipment.

In Fig. A.3, a (100) silicon wafer is referenced to its unity cube where are represented the three principal planes: (100), (110) and (100). It can be seen from this figure that intersections of the {111} planes with the {100} planes (the wafer surface) are mutually perpendicular and lying along the ⟨110⟩ orientations. This representation can be very helpful to understand the intersections of the main crystallographic planes and directions in the bulk of a silicon wafer. Angles between two planes (h_1, k_1, l_1) and (h_2, k_2, l_2) or two directions $[h_1, k_1, l_1]$ and $[h_2, k_2, l_2]$ is given by

$$\theta = \cos^{-1}\left[\frac{h_1 h_2 + k_1 k_2 + l_1 l_2}{\sqrt{(h_1^2 + k_1^2 + l_1^2)(h_2^2 + k_2^2 + l_2^2)}}\right] \tag{A.2}$$

Angles between [100] and [111] directions can be found as $\cos^{-1}\left(1/\sqrt{3}\right) = 54.74°$ or

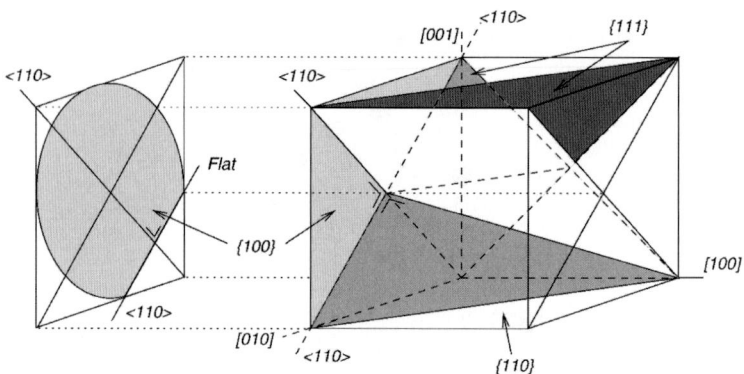

Figure A.3: (100) silicon wafer with reference to the unity cube and its relevant planes (After [27]).

$\cos^{-1}\left(-1/\sqrt{3}\right) = 125.26°$. Angles between [110] and [111] directions can be found as $\cos^{-1}\left(2/\sqrt{6}\right) = 35.26°$, $\cos^{-1}(0) = 90°$ and $\cos^{-1}\left(-2/\sqrt{6}\right) = 144.74°$.

Appendix B

About Interferometry...

From [80],[201],[81],[82] and [202].

The profilometer used to measure or detect the buckling of strain microstructures is based on light interferometry. It is an imaging set-up using a Michelson interferometer in a Linnik configuration (i.e. both arms of the interferometer have an identical microscope objective). The light source is a Light Emitting Diode (LED) and the detector is a CCD camera linked to a frame grabber. The LED has a spectral wavelength centered at 633 nm and low coherence (about 20 μm). Fig. B.1 represents this optical set-up.

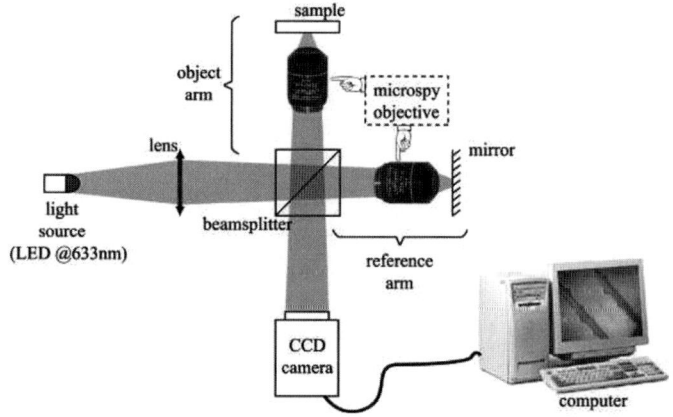

Figure B.1: Optical set-up.

The intensity measured by each pixel of the CCD detector is given by

$$I_p(X,Y) = \frac{I_0(x,y)}{2}[1 + V(x,y)\cos(\phi(x,y))] \quad (B.1)$$

where (X,Y) and (x,y) respectively represent the coordinates in the CCD system and the sample system, knowing that (X,Y) and (x,y) are related by the optical magnification; Ip represents the light intensity on the pixel located at coordinate (X,Y); I_0 represents the sum of the light intensities coming from the sample and the reference mirror and located at the (x,y) coordinates; $V(x,y)$ is the contrast of interferences at each sample point (x,y) and $\phi(x,y)$ is the phase of interferences which represents the optical path difference between interferometer arms. The latter can also be written as

$$\phi(x,y) = \frac{4\pi}{\lambda}\Delta z(x,y) \quad (B.2)$$

where Δz is the shape of the sample. More precisely, it represents the difference in length between the reference arm (ended by the mirror) and the object arm (ended by the sample). As we are considering the mirror as a flat reference, the resultant phase is an image of the sample profile.

Therefore, interferences recorded by the CCD camera can be used to obtain an image of the sample profile. In order to measure this profile, we need to extract the phase $\phi(x,y)$ inside the cosine term of Eq. B.1. This is done with a well-known technique named phase-shifting. This consists in recording four consecutive images (I_{p1}, I_{p2}, I_{p3} and I_{p4}), each image being phase shifted to each other. This is done by moving the reference mirror by a constant and calibrated length corresponding to a $\frac{\pi}{2}$ phase. The first image in Fig. B.2 represents an interferogram of an array of thermal silicon oxide clamped-clamped beams.

The phase can then be extracted by a combination of the four images:

$$\phi = \arctan\frac{I_{p2} - I_{p4}}{I_{p3} - I_{p1}} \quad (B.3)$$

As the algorithm uses an inverse tangent function, the phase is known modulo π but $I_{p2} - I_{p4}$ and $I_{p3} - I_{p1}$ are respectively proportional to $\sin(\phi)$ and $\cos(\phi)$, and a test on the sign is used in order to determine the quadrant of the phase. Therefore, the phase ϕ can be extracted modulo 2π (Fig. B.2(1) and (2)).

The 2π discontinuities can be removed by unwrapping techniques. The result is a continuous phase (Fig. B.2(3)) that is related to the shape by Eq. B.2. Finally, considering that $\Delta z = \frac{4\pi}{\lambda}\phi$, the profile of the buckling structure can be extracted (Fig. B.2(4)).

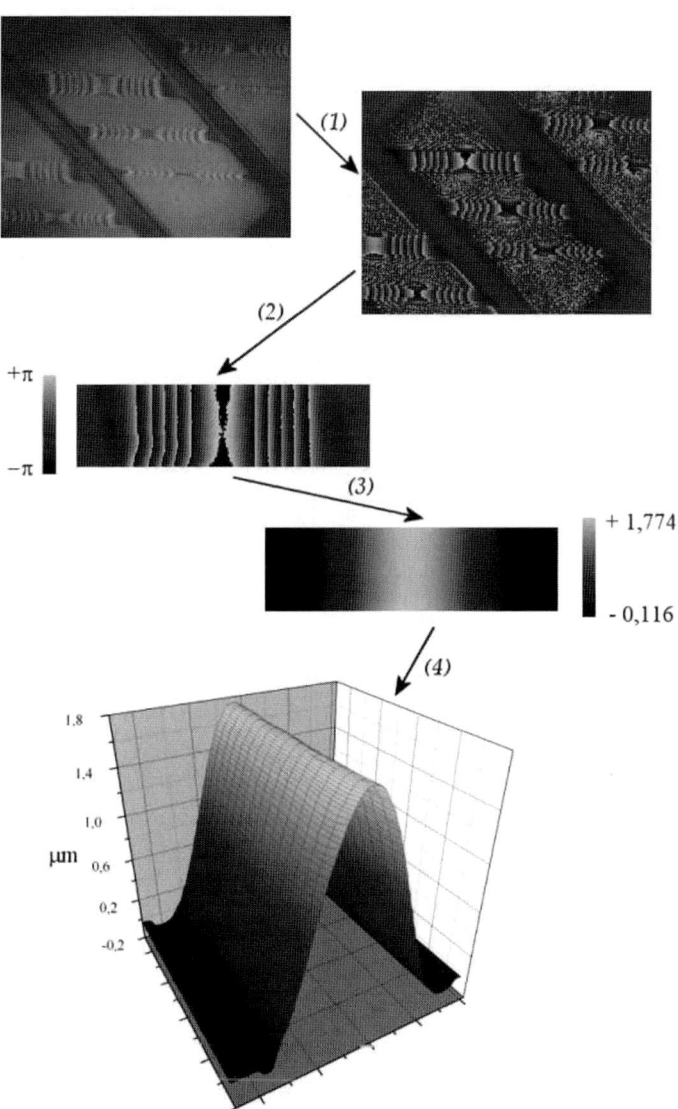

Figure B.2: The four steps to extract the profile of a clamped-clamped beam: (1) extraction of the phase from the interferogram; (2) selection of a part of the image; (3) unwrapping to give the continuous phase and (4) transformation in three-dimensional profile.

Appendix C

About Reflectometry...

From [203] and [204].

The temperature measurements on our heated membrane were based on the imaging reflectometry technique. The principle is based on the measurements of the relative variation of the light reflected by the device. In our case, the device is a membrane heated by a microheater and covered with additionnal polysilicon plots in order to increase its thermoreflectivity. Fig. C.1 summarizes the technique.

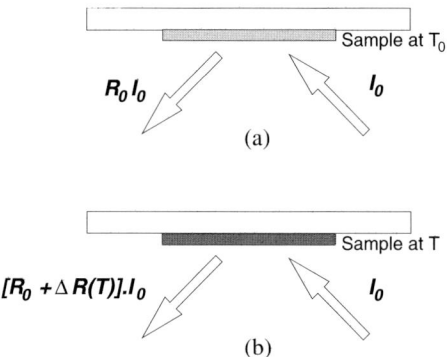

Figure C.1: Schematic of the light reflected by the device: (a) sample in standby at room temperature T_0 and (b), sample heated at temperature T.

Firstly, consider that a Light Emitting Diode (LED) lights the membrane (the polysilicon) in standby at room temperature T_0. The amount of reflected light I_{T0} by the membrane can be written,

$$I_{T0} = I_0 * R_0 \tag{C.1}$$

where I_0 is the incident light intensity and R_0, the reflection coefficient of the sample (so-called reflectivity).

Then, if we apply a current on the microheater in order to heat the membrane at a temperature T, the reflectivity $R(T)$ of the polysilicon will change according to its temperature,

$$R(T) = R_0 + \Delta R(T) = R_0 + \frac{\partial R}{\partial T}\Delta T \tag{C.2}$$

where $\Delta R(T)$ expresses the change of reflectivity versus the temperature T, $\frac{\partial R}{\partial T}$ is the reflectivity variation versus the temperature and, ΔT is the temperature variation reported to the room temperature. The light intensity I_T can therefore be written,

$$I_T = R(T) * I_0 = \left(R_0 + \frac{\partial R}{\partial T}\Delta T\right) * I_0 \tag{C.3}$$

Knowing I_T and I_{T0}, it is possible to calculate the temperature variation ΔT,

$$\Delta T = \left(\frac{1}{R_0}\frac{\partial R}{\partial T}\right)^{-1} * \frac{I_T - I_{T0}}{I_{T0}} = \kappa^{-1}\frac{I_T - I_{T0}}{I_{T0}} \tag{C.4}$$

with $\kappa = \frac{1}{R_0}\frac{\partial R}{\partial T}$, the so-called thermoreflectivity coefficient.

Consequently, we see it is possible to extract the variation of temperature ΔT (Eq. C.4) if we know the thermoreflectivity coefficient. To find its value, we need to calibrate previously the microheater, such as for its use as thermometer. Therefore, the microheater was firstly heated at a calibrated temperature in order to extract the thermoreflectivity coefficient.

Bibliography

[1] L. Ristic, *Sensor Technology and devices*. Boston: L.J. Ristic, Artech House ed., 1994.

[2] H. Baltes and O. Brand, "CMOS-based microsensors," *Sensors and Actuators A*, vol. 92, pp. 1–9, 2001.

[3] F. Udrea and J. Gardner, "SOI CMOS gas sensors," in *Proceeding of the First IEEE Sensors conference*, (Orlando, Florida, USA), pp. 1379–1384, IEEE, June 2002.

[4] K. Derbyshire, "As devices shrink and manufacturing costs drop, silicon-on-insulator's future is now," *Semiconductor Manufacturing, 7 pages*, December 2003.

[5] F. Foulon, L. Rousseau, L. Babadjian, S. Spirkovitch, A. Brambilla, and P. Bergonzo, "A new technique for the fabrication of thin silicon radiation detectors," *IEEE Transactions on nuclear science*, vol. 46, pp. 218–220, June 1999.

[6] D. Flandre, S. Adriaensen, A. Afzalian, J. Laconte, D. Levacq, C. Renaux, L. Vancaillie, and J.-P. Raskin, "Intelligent SOI CMOS integrated circuits and sensors for heterogeneous environments and applications," in *IEEE Sensors Conference*, (Orlando (Florida)), pp. 1407–1412, June 12-14 2002. Invited paper.

[7] D. Flandre, S. Adriaensen, A. Akheyar, A. Crahay, L. Demes, P. Delatte, V. Dessard, B. Iniguez, A. Nve, B. Katschmarskyj, P. Loumaye, J. Laconte, I. Martinez, G. Picun, E. Rauly, C. Renaux, D. Spte, M. Zitout, M. Dehan, B. Parvais, P. Simon, D. Vanhoenacker, and J. P. Raskin, "Fully depleted SOI CMOS technology for heterogenous micropower, high-temperature or RF microsystems," *Solid State Electronics, from Elsevier Science, Pergamon*, vol. 45, pp. 541–549, April 2001.

[8] V. Dessard, *SOI specific analog techniques for low-noise, high-temperature or ultra-low power circuits*. PhD thesis, Universit catholique de Louvain, Louvain-La-Neuve, Belgium, April 2001.

[9] B. Diem, P. Rey, S. Renard, S. V. Bosson, H. Bono, F. Michel, M. Delaye, and G. Delapierre, "SOI SIMOX; from bulk to surface micromachining, a new age for silicon sensors and actuators," *Sensors and Actuators A*, vol. 46-47, pp. 8–16, 1995.

[10] W. Mokwa, "Advanced sensors and microsystems on SOI," *Int. Journal of High Speed Electronics and Systems*, vol. 10, no. 1, pp. 147–153, 2000.

[11] G. T. Kovacs, *Micromachined Transducers Sourcebook*. USA: WCB/McGraw-Hill, February 1998.

[12] A. Mller and S. Iordanescu, "Membrane supported microwave circuits," in *CAS 1998 international semiconductor conference (21st edition)*, (Sinaia (Romania)), pp. 437–444, IEEE, 6-10 October 1998.

[13] C. Mastrangelo and W. Tang, "Semiconductor sensor technologies," in *Semiconductor Sensors* (S. Sze, ed.), ch. 2, pp. 17–95, USA: John Wiley and Sons, Inc, 1994.

[14] *HDMicrosystems, Pyralin PI2720 Processing Guidelines*.

[15] I. Petrini, "Microwave micromachined structures on ¡100¿ silicon substrate using polyimide supported pads," *Microelectronic Engineering*, vol. 51-52, pp. 595–600, 2000.

[16] M. Aslam and J. Hatfield, "Fabrication of thin film microheater for gas sensors on polyimide membrane," in *Proceedings of the 2nd IEEE Sensors Conference*, (Toronto, Canada), pp. 389–392, October 2003.

[17] J. Laconte, V. Wilmart, and J.-P. Raskin, "High-sensitivity capacitive humidity sensor using three-layer patterned polyimide sensing film," in *Proceedings of the 2nd IEEE Sensors Conference*, (Toronto, Canada), pp. 372–377, October 2003.

[18] S. Alcantara, A. Cerdeira, and G.Romero-Paredes, "MOS transistor pressure sensor," in *Proceedings of the Second IEEE International Caracas Conference on Devices, Circuits and Systems*, (Caracas), pp. 381–385, March 2-4 1998.

[19] M. Horrillo, I. Sayago, L. Ars, J. Rodriguo, J. Gutirrez, A. Gtz, I. Grcia, L. Fonseca, C. Can, and E. Lora-Tamayo, "Detection of low NO_2 concentrations with low power micromachined tin oxide gas sensors," *Sensors and Actuators B*, vol. 58, pp. 325–329, 1999.

[20] P. Ivanov, *Design, Fabrication and characterization of thick-film gas sensors*. PhD thesis, Universitat Rovira i Virgili, Tarragona, Spain, June 2004.

[21] O. Tabata, R. Asahi, H. Funabashi, K. Shimaoka, and S. Sugiyama, "Anisotropic etching of silicon in TMAH solutions," *Sensors and Actuators A*, vol. 34, pp. 51–57, 1992.

[22] E. H. Klaassen, *Micromachined Instrumentation systems*. PhD thesis, Stanford University, May 1996.

[23] O. Tabata, "pH controlled TMAH etchants for silicon micromachining," *Sensors and Actuators A*, vol. 53, pp. 335–339, 1996.

[24] E. H. Klaassen, R. J. Reay, and G. T. Kovacs, "Diode-based thermal r.m.s. converter with on-chip circuitry fabricated using CMOS technology," *Sensors and Actuators A*, vol. 52, pp. 33–40, 1996.

[25] G. T. Kovacs, N. I. Maluf, and K. E. Petersen, "Bulk micromachining of silicon," *Proceedings of the IEEE Integrated sensors, microactuators and microsystems (MEMS)*, vol. 86, pp. 1536–1551, August 1998.

[26] D. Lapadatu and H. Jakobsen, "Building of silicon mechanical sensors by bulk micromachining and anodic bonding," in *CAS 1998 international semiconductor conference (21st edition)*, (Sinaia (Romania)), pp. 33–44, IEEE, 6-10 October 1998.

[27] M. Madou, *Fundamentals of Microfabrication*. USA: CRC Press, 1997.

[28] K. E. Petersen, "Silicon as a mechanical material," *Proceedings of the IEEE*, vol. 70, pp. 420–457, May 1982.

[29] D. Veychard, *Conception et ralisation d'un convertisseur lectro-thermique grande constante de temps en technologie microsystme pour un disjoncteur thermique*. PhD thesis, Laboratoire TIMA, December 1999.

[30] E. H. Klaassen, R. J. Reasy, C. Storment, and G. T. Kovacs, "Micromachined thermally isolated circuits," *Sensors and Actuators A*, vol. 58, pp. 43–50, 1997.

[31] C. Moldovan, R. Iosub, G. Nechifor, D. Dascalu, F. Craciunoiu, and B. Servan, "The mechanism of anisotropic etching of silicon in a complexant alkaline system," in *CAS 1998 international semiconductor conference (21st edition)*, (Sinaia (Romania)), pp. 353–356, IEEE, 6-10 October 1998.

[32] J. Thong, W. Choi, and C. Chong, "TMAH etching of silicon and the interaction of etching parameters," *Sensors and Actuators A*, vol. 63, pp. 243–249, 1997.

[33] K. Sato, M. Shikida, T. Yamashiro, K. Asaumi, Y. Iriye, and M. Yamamoto, "Anisotropic etching rates of single-crystal silicon for TMAH water solution as a function of crystallographic orientation," *Sensors and Actuators A*, vol. 73, pp. 131–137, 1999.

[34] R. Mller, P. Obreja, E. Manea, and N. Nastase, "An investigation of silicon thin membranes for MOMS," in *CAS 1998 international semiconductor conference (21st edition)*, (Sinaia, Romania), pp. 345–348, IEEE, 6-10 October 1998.

[35] W. Choi, J. Thong, P. Luo, C. Tan, T. Chua, and Y. Bai, "Characterisation of pyramid formation arising from the TMAH etching of silicon," *Sensors and Actuators A*, vol. 71, pp. 238–243, 1998.

[36] J. S. You, D. Kim, J. Y. Huh, H. J. Park, J. J. Pak, and C. S. Kang, "Experiments on anisotropic etching of Si in TMAH," *Solar Energy Materials and Solar Cells*, vol. 66, pp. 37–44, 2001.

[37] J. Tong, Y. Bai, P. Luo, and W. Choi, "Investigations on the morphology of silicon surfaces anisotropically etched with TMAH," *Material Science and Engineering B*, vol. 72, pp. 177–179, 2000.

[38] M. Shikida, T. Masuda, D. Uchikawa, and K. Sato, "Surface roughness of single-crystal silicon etched by TMAH solution," *Sensors and Actuators A*, vol. 90, pp. 223–231, 2001.

[39] J. jye Tsaur, C.-H. Du, and C. Lee, "Investigation of TMAH for front-side bulk micromachining process from manufacturing aspect," *Sensors and Actuators A*, vol. 92, pp. 375–383, 2001.

[40] P.-H. Chen, H.-Y. Peng, C.-M. Hsieh, and M. K. Chyu, "The characteristic behavior of TMAH water solution for anisotropic etching of both silicon substrate and SiO_2 layer," *Sensors and Actuators A*, vol. 93, pp. 132–137, 2001.

[41] G. Yan, P. C. Chan, I.-M. Hsing, R. K. Sharma, J. K.O., and Y. Wang, "An improved TMAH Si-etching solution without attacking exposed aluminum," *Sensors and Actuators A*, vol. 89, pp. 135–141, 2001.

[42] K. Lian, B. Stark, A. Gundlach, and A. Walton, "Aluminum passivation for TMAH based anisotropic etching for MEMS applications," *Electronics Letters*, vol. 35, pp. 1266–1267, July 1999.

[43] S. Brida, A. Faes, V. Guarnieri, F. Giacomozzi, B. Margesin, M. Paranjape, G. Pignatel, and M. Zen, "Microstructures etched in doped TMAH solutions," *Microelectronic Engineering*, vol. 53, pp. 547–551, 2000.

[44] P. Sarro, D. Brida, W. Vlist, and S. Brida, "Effect of surfactant on surface quality of silicon microstructures etched in saturated TMAHW solutions," *Sensors and Actuators A*, vol. 85, pp. 340–345, 2000.

[45] M. Lambrechts and W. Sansen, "Micromachining techniques in silicon," in *Biosensors microelectrochemical devices*, ch. 4.3, pp. 118–131, New-York: Institute of Physics Publishing, 1992.

[46] E. Steinsland, M. Nese, A. Hanneborg, R. W. Bernstein, H. Sandmo, and G. Kittilsland, "Boron etch-stop in TMAH solutions," *Sensors and Actuators A*, vol. 54, pp. 728–732, 1996.

[47] R. Charavel, J. Laconte, and J.-P. Raskin, "Advantages of p++ polysilicon etch stop layer versus p++ silicon," in *Smart Sensors, Actuators, and MEMS Conference* (C. C. J.C. Chiao, V.K. Varadan, ed.), vol. 5116 part 2, (Gran Canaria, Spain), pp. 699–709, SPIE, 19-21 May 2003.

[48] B. Kloeck, S. D. Collins, N. F. D. Rooij, and R. L. Smith, "Study of electrochemical etch-stop for high-precision thickness control of silicon membranes," *IEEE Transaction on Electron Devices*, vol. 36, pp. 663–669, April 1989.

[49] M. Acero, J. Esteve, C. Burrer, and A. Gtz, "Electrochemical etch-stop characteristics of TMAH:IPA solutions," *Sensors and Actuators A*, vol. 46-47, pp. 22–26, 1995.

[50] R. L. Johnson, "High temperature Silicon-On-Insulator pressure sensor technology," in *The Third European Conference on High Temperature Electronics HITEN*, pp. 45–48, 1999.

[51] P. French, M. Nagao, and M. Esashi, "Electrochemical etch-stop in TMAH without externally applied bias," *Sensors and Actuators A*, vol. 56, pp. 279–280, 1996.

[52] C. Ashruf, P. French, P. C.Bressers, P. Sarro, and J. Kelly, "A new contactless electrochemical etch-stop based on a gold/silicon/TMAH galvanic cell," *Sensors and Actuators A*, vol. 66, pp. 284–291, 1998.

[53] C. Tellier, "CAD design of mask compensation patterns," in *Proceeding of the IEEE Sensors conference*, vol. 1, (Toronto, Canada), pp. 517–522, October 2003.

[54] N. D. Masters, M. P. de Boer, B. D. Jensen, M. S. Baker, and D. Koester, "Side-by-side comparison of passive MEMS strain test structures under residual compression," in *Mechanical Properties of Structural Films* (C. L. Muhlstein and S. B. Brown, eds.), vol. ASTM Special Technical Publication, 1413, pp. 1–33, American Society for Testing and Materials, October 2001.

[55] L. Lin, A. P. Pisano, and R. T. Howe, "A micro strain gauge with mechanical amplifier," *Journal of Microelectromechanical Systems*, vol. 6, pp. 313–321, December 1997.

[56] C. S. Pan and W. Hsu, "A microstructure for in situ determination of residual strain," *Journal of Microelectromechanical Systems*, vol. 8, pp. 200–207, June 1999.

[57] M. W. Frtsch, *Mechanical properties of thick polycrystalline silicon films suitable for surface micromachining*. PhD thesis, Gerhard-Mercator-Universitt-Gesamthochschule, Duisburg, 1998.

[58] O. Solgaard, "Chapter 7 stress and strain." Syllabus of course MEMS Design, Stanford University CA, May 2003.

[59] M. Ohring, *Material Science of Thin Films, Deposition and Structure; Chapter 12: Mechanical Properties of Thin Films*. USA: Academic Press, second ed., 2002.

[60] W. A. Brantley, "Calculated elastic constants for stress problems associated with semiconductor devices," *Journal of Applied Physics*, vol. 44, no. 1, pp. 534–535, 1973.

[61] D. R. Frana and A. Blouin, "All-optical measurement of in-plane and out-of-plane Young's modulus and Poisson ratio in silicon wafers by means of vibration modes," *Measurement Science and Technology*, vol. 15, pp. 859–868, March 2004.

[62] B. Stark, "MEMS reliability assurance guidelines for space applications," Jet Propulsion Laboratory Publication, NASA, California Institute of Technology, Pasadena, California, January 1999.

[63] B. Bhushan and X. Li, "Micromechanical and tribological characterization of doped single-crystal silicon and polysilicon films for microelectromechanical systems devices," *Journal of Materials Research*, vol. 12, pp. 54–63, January 1997.

[64] V. Domnich and Y. Gogotsi, "Phase transformations in silicon under contact loading," *Rev. Adv. Mater. Sci*, vol. 3, pp. 1–36, 2002.

[65] N. Hossain, J. Ju, B. Warneke, and K. Pister, "Characterization of the Young's modulus of CMOS thin films," in *Mechanical Properties of Structural Films* (C. L. Muhlstein and S. B. Brown, eds.), vol. ASTM Special Technical Publication, 1413, pp. 1–13, ASTM Special Technical Publication, 1413, July 2001.

[66] D. Herman, M. Gaitan, and D. D. Voe, "MEMS test structures for mechanical characterization of VLSI thin films," in *SEM Conference*, (Portland, Oregon), p. 5, June 4-6 2001.

[67] H. A. Rueda and M. E. Law, "Modeling of strain in boron-doped silicon cantilevers," in *International Conference on Modeling and Simulations of Microsystems (MSM)*, vol. 1, (USA), pp. 94–99, 1998.

[68] W. Fang and J. Wickert, "Comments on measuring thin-film stresses using bi-layer micromachined beams," *Journal of Micromechanics and Microengineering*, vol. 5, pp. 276–281, 1995.

[69] O. Solgaard, "Chapter 8 mechanical structures." Syllabus of course MEMS Design, Stanford University CA, May 2003.

[70] C. A. Klein, "How accurate are Stoney's equation and recent modifications," *Journal of Applied Physics*, vol. 88, pp. 5487–5489, November 2000.

[71] P. Townsend, D. Barnett, and T. Brunner, "Elastic relationships in layered composite media with approximation for the case of thin films on a thick substrate," *Journal of Applied Physics*, vol. 62, pp. 4438–4444, December 1987.

[72] R. Charavel, B. Olbrechts, and J.-P. Raskin, "Stress release of PECVD oxide by RTA," in *Smart Sensors, Actuators, and MEMS Conference* (C. C. J.C. Chiao, V.K. Varadan, ed.), vol. 5116 part 2, (Gran Canaria, Spain), pp. 596–606, SPIE, 19-21 May 2003.

[73] C. Rossi, P. Temple-Boyer, and D. Estve, "Realization and performance of thin SiO_2-SiN_x membrane for microheater applications," *Sensors and Actuators A*, vol. 64, pp. 241–245, 1998.

[74] B. C. Chou, J.-S. Shie, and C.-N. Chen, "Fabrication of low-stress dielectric thin-film for microsensor applications," *IEEE Electron Device Letters*, vol. 18, pp. 599–601, December 1997.

[75] P. Temple-Boyer, B. Hajji, J. Alay, J. Morante, and A. Martinez, "Properties of SiO_xN_y films deposited by LPCVD from $SiH_4/N_2O/NH_3$ gaseous mixture," *Sensors and Actuators A*, no. 74, pp. 52–55, 1999.

[76] S. Asti, A. Gu, E. Scheid, and J. Guillemet, "Design of a low power SnO_2 gas sensor integrated on silicon oxynitride membrane," *Sensors and Actuators B*, vol. 67, pp. 84–88, 2000.

[77] W. Fang and J. Wickert, "Determining mean and gradient residual stresses in thin films using micromachined cantilevers," *Journal of Micromechanics and Microengineering*, vol. 6, pp. 301–309, 1996.

[78] W. Fang and J. Wickert, "Post buckling of micromachined beams," *Journal of Micromechanics and Microengineering*, vol. 4, pp. 116–122, 1994.

[79] S. A. Smee, M. Gaitan, D. B. Novotny, Y. Joshi, and D. L. Blackburn, "IC test structures for multilayer interconnect stress determination," *IEEE Electron Device Letters*, vol. 21, January 2000.

[80] S. Jorez, *Dveloppement d'instrumentation et de mthodologies pour la caractrisation thermique et thermomcanique de composants lectroniques*. PhD thesis, Universit Bordeaux 1, Bordeaux, octobre 2001.

[81] Dubois, Boccara, and Lebec, "Real-time reflectivity and topography imagery of depth-resolved microscopic surfaces," *Optics Letters*, vol. 24, no. 5, pp. 309–311, 1999.

[82] S. Grauby, S. Dilhaire, S. Jorez, and W. Claeys, "Imaging set-up for temperature, topography and surface displacement measurements of microelectronic devices," *Review of Scientific Instruments*, vol. 74, no. 1, pp. 645–647, 2003.

[83] H. Guckel, D. Burns, C. Rutigliano, E. Lovell, and B. Choi, "Diagnostic microstructures for the measurement of intrinsic strain in thin films," *Journal of Micromechanical Microengineering*, vol. 2, p. 86, 1992.

[84] T. V. der Donck, J. Proost, C. Rusu, K. Baert, C. V. Hoof, J.-P. Celis, and A. Witvrouw, "Effect of deposition parameters on the stress gradient of CVD and PECVD poly-SiGe for MEMS applications," in *Proceedings of the SPIE, Conference on Micromachining and Microfabrication Process Technology*, vol. 5342, pp. 8–18, 2004.

[85] B. Parvais, A. Pallandre, A. Jonas, and J.-P. Raskin, "A fluoro-ethoxysilane-based stiction-free release process for submicron gap MEMS," in *NanoTech 2003 conference*, vol. 1, (San Francisco CA), pp. 522–525, February 2003.

[86] I. H. Jafri, H. D. Moritz, H. H. Busta, and S. Walsh, "Supercritical carbon dioxide drying and cleaning: Application to MEMS technology," *MSTnews international newsletter on Microsystems and MEMS*, vol. 2, pp. 15–20, 1999.

[87] A. Witvrouw, B. D. Bois, P. D. Moor, A. Verbist, C. V. Hoof, H. Bender, and K. Baert, "A comparison between wet HF etching and vapor HF etching for sacrificial oxide removal," in *Proc. SPIE Micromachining and Microfabrication Process Technology VI*, vol. 4174, pp. 130–141, SPIE, Sept. 2000.

[88] D. Briand, M.-A. Grtillat, B. van der Schoot, and N. de Rooij, "Thermal management of micro-hotplates using MEMCAD as simulation tool," in *3rd International Conference on Modelling and Simulation of Microsystems*, (San Diego, California, USA), pp. 640–643, IEEE, March 27-29 2000.

[89] J. Suehle, R. Cavicchi, M. Gaitan, and S. Semancik, "Tin oxide gas sensor fabricated using CMOS micro-hotplates and in-situ processing," *IEEE Electron Device Letters*, vol. 14, pp. 118–120, March 1993.

[90] T. Kunt, T. McAvoy, R. Cavicchi, and S. Semancik, "Optimization of temperature programmed sensing for gas identification using micro-hotplate sensors," *Sensors and Actuators B*, vol. 53, pp. 24–43, 1998.

[91] S. Semancik, R. Cavicchia, M. Wheeler, J. Tiffany, G. Poirier, R. Walton, J. Suehle, B. Panchapakesan, and D. D. Voe, "Microhotplate platforms for chemical sensor research," *Sensors and Actuators B*, vol. 77, pp. 579–591, 2001.

[92] M. Afridi, J. Suehle, M. Saghloul, J. Tiffany, and R. Cavicchi, "Implementation of CMOS compatible conductance-based micro-gas-sensor system," in *European Conference on Circuit Theory and Design*, vol. 3, (Espoo, Finland), pp. 381–384, August 28-31 2001.

[93] M. Afridi, J. Suehle, M. Zaghloul, D. Berning, A. Hefner, R. Cavicchi, S. Semancik, C. Montgomery, and C. Taylor, "A monolithic CMOS microhotplate-based gas sensor system," *IEEE Sensors Journal*, vol. 2, pp. 644–655, December 2002.

[94] Y. Mo, Y. Okawa, K. Inoue, and K. Natukawa, "Low-voltage and low-power optimization of micro-heater and its on-chip drive circuitry for gas sensor array," *Sensors and Actuators A*, vol. 100, pp. 94–101, 2002.

[95] D. Barrettino, M. Graf, M. Zimmermann, C. Hagleitner, A. Hierlemann, and H. Baltes, "A smart single-chip micro-hotplate-based gas sensor system in CMOS-technology," *Analog Integrated Circuits and Signal Processing*, vol. 39, pp. 275–287, 2004.

[96] P. Roetsch, H. Bttner, K. Seibert, A. Gtz, I. Grcia, J. Plaza, and C. Can, "Combination of metal oxyde based gas sensors with CMOS electronic circuits on wafer level," in *Eurosensors XIV The 14th European Conference on Solid-State Transducers*, (Copenhagen, Denmark), pp. 377–378, August 27-30 2000.

[97] A. Gtz, I. Grcia, C. Can, M. Lozano, and E. Lora-Tamayo, "Thermo-mechanical structures for the optimisation of silicon micromachined gas sensors," in *ICMTS International Conference on Microelectronic Test Structures*, (Monterey, USA), pp. 90–94, IEEE, March 2000.

[98] V. Guidi, G. Cardinali, L. Dori, G. Faglia, M. Ferroni, G. Martinelli, P. Nelli, and G. Sberveglieri, "Thin-film gas sensor implemented on a low-power-consumption micromachined silicon structure," *Sensors and Actuators B*, vol. 49, pp. 88–92, 1998.

[99] D. Briand, A. Krauss, B. van der Schoot, U. Weimar, N. Barsan, W. Gpel, and N. de Rooij, "Design and fabrication of high-temperature micro-hotplates for drop coated gas sensors," *Sensors and Actuators B*, vol. 68, pp. 223–233, 2000.

[100] A. Gtz, I. Grcia, C. Can, E. Lora-Tamayo, M. Horrillo, J. Getino, C. Garcia, and J. Guttirrez, "A micromachined solid state integrated gas sensor for the detection of aromatic hydrocarbons," *Sensors and Actuators B*, vol. 44, pp. 483–487, 1997.

[101] A. Tinoco, E. Llobet, J. Brezmes, M. Stankova, P. Ivanov, X. Vilanova, X. Correig, I. Grcia, and C. Can, "MLS based temperature modulation of micro-hotplates," in *Proceedings of the 2nd IEEE Sensors Conference*, (Toronto, Canada), pp. 1255–1259, IEEE, October 2003.

[102] N. Das, C. Monroy, D. Robinson, and M. Jhabvala, "Design and fabrication of low power polysilicon sources," *Solid-State Electronics*, vol. 43, pp. 1239–1244, 1999.

[103] O. Grudin, G. Frolov, I. Katsan, and B. Lupina, "Thermal microsensor with a.c. heating for gas-pressure measurements," *Sensors and Actuators A*, vol. 62, pp. 571–575, 1997.

[104] C. Kuratli and Q. Huang, "Ultrasound range-finder microsystem," *IEEE Journal of Solid State circuits*, vol. 35, pp. 2005–2017, December 2000.

[105] C. Can, I. Grcia, A. Gtz, J. Cazalla, M. Duch, M. Horrillo, I. Sayago, and J. Robla, "Electrical characterisation of polysilicon heaters for micromachined gas sensors," in *CDE 2a Conferencia de dispositivos electronicos*, (Madrid), pp. 13–16, June 1999.

[106] D.-D. Lee, W.-Y. Chung, M.-S. Choi, and J.-M. Baek, "Low-power micro gas sensor," *Sensors and Actuators B*, vol. 33, pp. 147–150, 1996.

[107] W.-Y. Chung, C.-H. Shim, and D.-D. Lee, "Tin oxide microsensor for LPG monitoring," *Sensors and Actuators B*, vol. 20, June 1994.

[108] M. Ehmann, F. Schubert, P. Ruther, and O. Paul, "Thermally activated ageing of polysilicon," in *IEEE Sensors conference*, vol. 1, (Orlando, Florida, USA), pp. 602–606, June 2002.

[109] M. Ehmann, P. Ruther, M. von Arx, H. Baltes, and O. Paul, "Ageing behavior of polysilicon heaters for CMOS microstructures operated at temperatures up to 1200K," in *The 14th IEEE International Conference MEMS*, pp. 147–150, January 21-25 2001.

[110] I. Graci, J. Santander, C. Can, M. Horrillo, I. Sayago, and J. Gutierrez, "Results on the reliability of silicon micromachined structures for semiconductor gas sensors," *Sensors and Actuators B*, vol. 77, pp. 409–415, 2001.

[111] X. Zhang, A. Mehra, A. Ayon, and I. Waitz, "Igniters and temperature sensors for a micro-scale combustion system," *Sensors and Actuators A*, vol. 103, pp. 253–262, 2003.

[112] M. von Arx, O.Paul, and H. Baltes, "Test structures to measure the heat capacity of CMOS layer sandwiches," *IEEE Transactions on Semiconductor manufacturing*, vol. 11, pp. 217–224, May 1998.

[113] M. von Arx, O. Paul, and H. Baltes, "Process-dependend thin film thermal conductivities for thermal CMOS MEMS," *Journal of Microelectromechanical Systems*, vol. 9, pp. 136–145, March 2000.

[114] S. Asti, A. Gu, E. Scheid, L. Lescouzres, and A. Cassagnes, "Optimization of an integrated SnO_2 gas sensor using a FEM simulator," *Sensors and Actuators A*, vol. 69, pp. 205–211, 1998.

[115] A. Gtz, I. Grcia, C. Can, and E. Lora-Tamayo, "Thermal and mechanical aspects for designing micromachined low-power gas sensors," *Journal of Micromechanics and Microengineering*, vol. 7, pp. 247–249, 1997.

[116] D. Briand, G.-M. Tomassone, and N. de Rooij, "Accelerated ageing of microhotplates for gas sening applications," in *Proceedings of the 2nd IEEE Sensors Conference*, (Toronto, Canada), pp. 1314–1317, October 2003.

[117] C. Can, I. Grcia, A. Gtz, L. Fonseca, E. Lora-Tamayo, M. Horrillo, I. Sayago, J. Robla, J. Rodriguo, and J. Gutirrez, "Detection of gases with arrays of micromachined tin oxide gas sensors," *Sensors and Actuators B*, vol. 65, pp. 244–246, 2000.

[118] S. Majoo, J. Gland, K. Wise, and J. Schwank, "A silicon micromachined conductometric gas sensor with a maskless Pt sensing film deposited by selected-area CVD," *Sensors and Actuators B*, vol. 35-36, pp. 312–319, 1996.

[119] P. Ruther, M. Ehmann, T. Lindermann, and O. Paul, "Dependence of the temperature distribution in micro hotplates on heater geometry and heating mode," in *12th International Conference on Transducers, Solid-State Sensors, Actuators and Microsystems*, vol. 1, (Boston), pp. 73–76, June 8-12 2003.

[120] M. Berger and Z. Chai, "Estimation of heat transfer in SOI-MOSFET's," *IEEE Transactions on Electron Devices*, vol. 38, pp. 871–875, April 1991.

[121] A. MConnell, S. Uma, and K. Goodson, "Thermal conductivity of doped polysilicon layers," *Journal of Microelectromechanical Systems*, vol. 10, pp. 360–369, 2001.

[122] A. Irace and P. Sarro, "Measurement of thermal conductivity and diffusivity of single and multilayer membranes," *Sensors and Actuators A*, vol. 76, pp. 323–328, 1999.

[123] C. Mastrangelo, J. H.-J. Yeh, and R. Muller, "Electrical and optical characteristics of vacuum sealed polysilicon microlamps," *IEEE Transactions on Electron Devices*, vol. 39, pp. 1363–1375, June 1992.

[124] E. Obermeier, P. Kopystynski, and R. Niebl, "Characteristics of polysilicon layers and their application in sensors," in *IEEE Solid-State Sensors Workshop*, (Hilton Head Is., SC), 1986.

[125] H.-M. Chuang, K.-B. Thei, S.-F. Tsai, and W.-C. Liu, "Temperature-dependent characteristics of polysilicon and diffused resistors," *IEEE Transactions on Electron Devices*, vol. 50, no. 5, pp. 1413–1415, 2003.

[126] J. Lee, "Class notes."

[127] D. Vicenzi, M. Butturi, M. Stefancich, C. Malag, V. Guidi, M. Carotta, G. Martinelli, V. Guarnieri, S. B. ans B. Margesin, F. Giacomozzi, M. Zen, A. Vasiliev, and A. Pisliakov, "Low-power thick-film gas sensor obtained by a combination of screen printing and micromachining techniques," *Thin Solid Films from Elesevier*, vol. 391, pp. 288–292, 2001.

[128] D. Vincenzi, M. Butturi, V. Guidi, M. Carotta, G. Martinelli, V. Guarnieri, S. Brida, B. Margesin, F. Giacomozzi, M. Zen, G. Pignatel, A. Vasiliev, and A. Pisliakov, "Development of a low-power thick-film gas sensor deposited by screen-printing technique onto a micromachined hotplate," *Sensors and Actuators B*, vol. 77, pp. 95–99, 2001.

[129] C. Tsamis, A. Nassiopoulou, and A. Tserepi, "Thermal properties of suspended porous silicon micro-hotplates for sensor applications," in *Eurosensors XVI The 16th European Conference on Solid-State Transducers*, (Prague, Czech Republic), September 15-18 2002.

[130] J. Covington, F. Udrea, and J. Gardner, "Resistive gas sensor with integrated MOSFET micro hot-plate based on an analogue SOI CMOS process," in *Proceeding of the First IEEE Sensors conference*, (Orlando, Florida, USA), IEEE, June 2002.

[131] J. Babcock, D. Feldbaumer, and V. Mercier, "Polysilicon resistor trimming for packaged integrated circuits," in *International Electron Device Meeting (IEDM)* (T. Digest, ed.), pp. 247–250, 1993.

[132] D. Feldbaumer, J. Babcock, V. Mercier, and C. Chun, "Pulse current trimming of polysilicon resistors," *IEEE Transactions on Electron Devices*, vol. 42, pp. 689–696, April 1995.

[133] D. Feldbaumer and J. Babcock, "Theory and application of polysilicon resistor trimming," *Solid-State Electronics*, vol. 38, no. 11, pp. 1861–1869, 1995.

[134] J. Babcock, P. Francis, R. Bashir, A. Kabir, D. Schroder, M. Lee, T. Dhayagude, W. Yindeepol, S. Prasad, A. Kalnitsky, M. Thomas, H. Haggag, K. Egan, A. Bergemont, and P. Jansen, "Precision electrical trimming of very low TCR poly-SiGe resistors," *IEEE Electron Device Letters*, vol. 21, pp. 283–285, June 2000.

[135] S. Das and S. Lahiri, "Electrical trimming of ion-beam-sputtered polysilicon resistors by high current pulses," *IEEE Transactions on Electron Devices*, vol. 41, pp. 1429–1434, August 1994.

[136] K. Kato and T. Ono, "Change in temperature coefficient of resistance of heavily doped polysilicon resistors caused by electrical trimming," *Japanese Journal of Applied Physics*, vol. 35, pp. 4209–4215, August 1996.

[137] M. Ehmann, P. Ruther, M. von Arx, and O. Paul, "Operation and short-term drift of polysilicon-heated CMOS microstructures at temperatures up to 1200 K," *Journal of Micromechanics and Microengineering*, vol. 11, pp. 397–401, July 2001.

[138] X. Zhang, A. Mehra, A. Ayon, and I. Waitz, "Development of polysilicon igniters and temperature sensors for a micro gas turbine engine," in *15th IEEE MEMS Conference*, (Las Vegas, USA), pp. 280–283, January 20-24 2002.

[139] N. Nguyen, "Micromachined flow sensors - a review," *Flow Meas. Instrum. from Elsevier*, vol. 8, no. 1, pp. 7–16, 1997.

[140] F. Kohl, R. Fasching, F. Keplinger, R. Chabicovsky, A. Jachimowicz, and G. Urban, "Development of miniaturized semiconductor flow sensors," *Measurement from Elsevier*, vol. 33, pp. 109–119, 2003.

[141] J. Schieferdecker, R. Quad, E. Holzenkmpfer, and M. Schulze, "Infrared thermopile sensors with high sensitivity and very low temperature coefficient," *Sensors and Actuators A*, vol. 46-47, pp. 422–427, 1995.

[142] H. Baltes, O. Paul, and O. Brand, "Micromachined thermally based CMOS microsensors," *Proceedings of the IEEE*, vol. 86, pp. 1660–1678, August 1998.

[143] D. Moser and H. Baltes, "A high sensitivity CMOS gas flow sensor on a thin dielectric membrane," *Sensors and Actuators A*, vol. 37-38, pp. 33–37, 1993.

[144] R. Reay, E. Klaassen, and G. Kovacs, "Thermally and electrically isolated single crystal silicon structures in CMOS technology," *IEEE Electron Device Letters*, vol. 15, pp. 399–401, October 1994.

[145] T. Akin, Z. Olgun, O. Akar, and H. Kulah, "An integrated thermopile structure with high responsivity using any standard CMOS process," *Sensors and Actuators A*, vol. 66, pp. 218–224, 1998.

[146] F. Mayer, A. Hberli, H. Jacobs, G. Ofner, O. Paul, and H. Baltes, "Single-chip CMOS anemometer," in *Technical Digest of the Electron Devices Meeting*, pp. 895–898, IEEE, December, 7-10 1997.

[147] M. Ashauer, H. Glosch, F. Hedrich, N. Hey, H. Sandmaier, and W. Lang, "Thermal flow sensor for liquids and gases based on combinations of two principles," *Sensors and Actuators A*, vol. 73, pp. 7–13, 1999.

[148] G. Kaltsas and A. Nassiopoulou, "Novel C-MOS compatible monolithic silicon gas flow sensor with porous silicon thermal isolation," *Sensors and Actuators A*, vol. 76, pp. 133–138, 1999.

[149] K. Makinwa and J. Huijsing, "A smart wind sensor using thermal sigma-delta modulation techniques," *Sensors and Actuators A*, vol. 97-98, pp. 15–20, 2002.

[150] I. Hariadi, H. Trieu, W. Mokwa, and H. Vogt, "Integrated flow sensor with monocrystalline silicon membrane operating in thermal time-of-flight mode," in *Proceedings of the 16th European Conference on Solid-State Transducers*, September 2002.

[151] R. Kersjes and W. Mokwa, "A fast liquid flow sensor with thermal isolation by oxide-filled trenches," *Sensors and Actuators A*, vol. 46-47, pp. 373–379, 1995.

[152] T. Neda, K. Nekamura, and T. Takumi, "A polysilicon flow sensor for gas flow meters," *Sensors and Actuators A*, vol. 54, pp. 626–631, 1996.

[153] S.-T. Hung, S.-C. Wong, and W. Fang, "The development and application of microthermal sensors with a mesh-membrane supporting structure," *Sensors and Actuators A*, vol. 84, pp. 70–75, 2000.

[154] R. Adamec, D. Thiel, and P. Tanner, "MEMS wind direction detection: from design to operation," in *Proceedings of the 2nd IEEE sensors conference*, (Toronto, Canada), pp. 340–343, October 2003.

[155] B. Rue and L. Otte, "Conception de circuits intgrs analogiques en technologie SOI soumis des hautes tempratures (300°C). Application: interfaage de MEMS (capteur de flux)," Master's thesis, Universit catholique de Louvain, Louvain-La-Neuve, Belgium, September 2003.

[156] U. Dillner, E. Kessler, S. Poser, V. Baier, and J. Mller, "Low power consumption thermal gas-flow sensor based on thermopiles of highly effective thermoelectric materials," *Sensors and Actuators A*, vol. 60, pp. 1–4, 1997.

[157] U. Dillner, E. Kessler, S. Poser, V. Braier, and J. Mller, "Thermal simulation of a micromachined thermopile-based thin-film gas flow sensor," *Microelectronics Journal*, vol. 29, pp. 291–297, 1998.

[158] P. Frjes, G. Lgradi, C. Dcso, A. Aszodi, and I. Barsony, "Modelling and characterisation of a micro gas-flow sensor," in *Proceedings of the 13th Micromechanics Europe Workshop (MME)*, (Sinaia), October 2002.

[159] T. Siewert, S. Liu, D. Smith, and J. Madeni, "Database for solder properties with emphasis on new lead-free solder," tech. rep., National Institute of Standards and Technology and Colorado School of Mines, Colorado, February 11 2002.

[160] P. Walker and W. Tarn, *Handbook of Metal etchants*. USA: P. Walker and W.H. Tarn, CRC press ed., 1991.

[161] S. Firebaugh, K. Jensen, and M. Schmidth, "Investigation of hight-temperature degradation of Platinum thin films with an in situ resistance measurement apparatus," *Journal of Microelectromechanical Systems*, vol. 7, pp. 128–135, March 1998.

[162] D. Rickerby, N. Wchter, M. Horrillo, J. Gutirrez, I. Grcia, and C. Can, "Structural and dimensional control in micromachined integrated solid state sensors," *Sensors and Actuators B*, vol. 69, pp. 314–319, 2000.

[163] J. Puigcorb, D. Vogel, B. Michel, A. Vil, I. Grcia, C. Can, and J. Morante, "High temperature degradation of Pt-Ti electrodes in microhotplate gas sensors," *Journal of micromechanics and Microengineering*, vol. 13, pp. S119–S124, July 2003.

[164] P. Ivanov, E. Llobet, X. Vilanova, M. Stankova, J. Bresmez, X. Correig, J. Hubalek, L. Malysz, I. Grcia, and C. Can, "Screen-printed nano-grain WO_3 films for micro-hotplates gas sensors," in *Proceedings of the 2nd IEEE Sensors Conference*, (Toronto, Canada), pp. 358–363, IEEE, October 2003.

[165] G. Faglia, E. Comini, A. Cristalli, G. Sberveglieri, and L. Dori, "Very low power consumption micromachined CO sensors," *Sensors and Actuators B*, vol. 55, pp. 140–146, 1999.

[166] P. Ivanov, M. Stankova, E. Llobet, X. Vilanova, I. Grcia, C. Can, and X. Correig, "Microhotplate sensor arrays based on sputtered and screen-printed metal oxide films for selective detection of volatile compounds," *Sensors and Transducers Magazine*, vol. 36, pp. 16–23, 2003.

[167] Y. Choi, G. Sakai, K. Shimanoe, N. Miura, and N. Yamazoe, "Wet process-prepared thick films of WO_3 for NO_2 sensing," *Sensors and Actuators B*, vol. 95, pp. 258–265, 2003.

[168] X. He, J. Li, X. Gao, and K. Wang, "NO_2 sensing characteristics of WO_3 thin film micro gas sensor," *Sensors and Actuators B*, vol. 93, pp. 463–467, 2003.

[169] Y. Zhao, Z. Feng, and Y. Liang, "Pulsed laser deposition of WO_3-based film for NO_2 gas sensor application," *Sensors and Actuators B*, vol. 66, pp. 171–173, 2000.

[170] G. Korotchenkov, V. Brynzari, and S. Dmitriev, "Electrical behaviour of SnO_2 thin films in humid atmosphere," *Sensors and Actuators B*, vol. 54, pp. 197–201, 1999.

[171] J.-P. Colinge, *Silicon-on-insulator technology: Materials to VLSI*. Kluwer, 2 ed., 1997.

[172] E. Sicard and S. Bendhia, *A book on deep submicron CMOS design using Microwind*. Brooks/Cole, in press ed., 2004.

[173] I. D. Houck, "Conception de cellules mmoire EEPROM en technologie CMOS SOI compltemement dplt," Master's thesis, Universit catholique de Louvain, Louvain-La-Neuve, Belgium, June 2004.

[174] F. van de Wiele and J.-P. Colinge, *Physique des dispositifs semi-conducteurs*. Paris, Bruxelles: De Boek–Wesmael, 1996.

[175] M. Bawedin and D. Izamova, "Transistor MOS submicronique," tech. rep., Universit catholique de Louvain, Louvain-La-Neuve, Belgium, 2003.

[176] M. Depas, B. Vermeire, P. Mertens, R. V. Meirhaeghe, and M. Heyns, "Determination of tunneling parameters in ultra-thin oxide layer poly-Si/SiO$_2$/Si structures," *Solid-State Electronics*, vol. 38, pp. 1465–1471, 1995.

[177] M. Bawedin, M. Estrada, and D. Flandre, "Investigation of floating body effects on SOI MOSFET gate tunneling currents," in *4th European Workshop on Ultimate Integration of Silicon*, pp. 45–48, March 2003.

[178] K. Honer, *Surface micromachining techniques for integrated microsystems*. PhD thesis, Stanford University, Stanford, Californy, USA, March 2001.

[179] S. An, S. Oh, and S. Kim, "CMOS compatibility of a micromachining process developed for semiconductor neural probe," in *Proceedings of the 23rd Annual EMBS International Conference* (IEEE, ed.), (Istambul, Turkey), pp. 3443–3445, October 25-28 2001.

[180] D. Flandre, F. Silveira, J.-P. Eggermont, B. Gentinne, V. Dessard, A. Viviani, D. Baldwin, L. Demes, and P. Jespers, "Design automation of CMOS OTAs using symbolic analysis and gm/ID methodology," in *SMACD conference*, (Leuven, Belgium), October, 10-11 1996.

[181] S. Wolf and R. N. Tauber, *Silicon Processing for the VLSI Era Vol. 1: Process Technology*, vol. 1. Sunset Beach, California, USA: Lattice Press, 1986.

[182] A. Tokuhiro and M. Bertino, "Tutorials on radiation effects on devices," tech. rep., Nuclear Engineering and Physics Department at the University of Missouri-Rolla, Missouri-Rolla, 2003.

[183] F. Anghinolfi, "Radiation hard electronics." Oral presentation, 2000.

[184] Z. Davis, G. Abadal, B. Helbo, O. Hansen, F. Campabadal, F. Prez-Murano, J. Esteve, E. Figueras, J. Verd, N. Barniol, and A. Boisen, "Monolithic integration of mass sensing nano-cantilevers with CMOS circuitry," *Sensors and Actuators A*, vol. 105, pp. 311–319, 2003.

[185] D. Brown, "Process training metal deposition (from the MIRC in georgia)." November 2003.

[186] J. Felix, J. Schwank, C. Cirba, R. Schrimpf, M. Shaneyfelt, D. Fleetwood, and P. Dodd, "Influence of total-dose radiation on the electrical characteristics of SOI MOSFETs," *Microelectronic Engineering*, vol. 72, pp. 332–341, 2004.

[187] A. Bullock, "SCI Team PDR Report," tech. rep., College of Engineering of the University of Hawaii, Manoa, Hawaii, March 2002.

[188] M. Renzi, M. Tate, A. Ercan, S. Gruner, E. Fontes, C. Powell, A. MacPhee, S. Narayanan, J. Wang, Y. Yue, and R. Cuenca, "Pixel array detectors for time resolved radiography," *Review of scientific instruments from IOP*, vol. 73, pp. 1621–1624, March 2002.

[189] A. Jordan, "Rad hard, space ready: Evolution of a fab-independent radiation-hardened COTS IC supplier," *COTS Journal*, p. 5 p., November 2001.

[190] W. Jenkins and S. Liu, "Radiation response of fully-depleted MOS transistors fabricated in SIMOX," *IEEE Transactions on nuclear science*, vol. 41, pp. 2317–2321, December 1994.

[191] T. Rudenko, A. Rudenko, V. Kilchytska, S. Crisoloveanu, T. Ernst, J.-P. Colinge, V. Dessard, and D. Flandre, "Determination of film and surface recombination in thin-film SOI devices using gated-diode technique," *Solid State Electronics*, vol. 48, pp. 389–399, 2004.

[192] Z. Savic and B. Radjenovic, "A method for separating the effects of interface from border and oxide trapped charge densities in MOS transistors," *Microelectronic Reliability*, vol. 37, no. 7, pp. 1147–1150, 1997.

[193] F. Rasmussen, M. Heschel, and O. Hansen, "Batch processing of CMOS compatible feedthroughs," *Microelectronic Engineering*, vol. 67-68, pp. 487–494, 2003.

[194] J. Pretet, S. Monfray, S. Cristoloveanu, and T. Skotnicki, "Silicon-on-Nothing MOS-FETs: performance, short-channel effects, and backgate coupling," *IEEE Transactions on Electron Devices*, vol. 51, pp. 240–245, February 2004.

[195] Z. Wang, J. Suski, D. Collard, and E. Dubois, "Piezoresistivity effects in n-MOSFET devices," *Sensors and Actuators A*, vol. 34, pp. 59–65, 1992.

[196] T. Akiyama, N. Blanc, and N. de Rooij, "A force sensor using a CMOS inverter in view of its application in scanning force microscopy," in *The Ninth Annual International Workshop on MEMS*, pp. 447–450, IEEE, February 11-15 1996.

[197] D. Briand, B. van der Schoot, N. de Rooij, H. Sundgren, and I. Lundstrm, "A low-power micromachined MOSFET gas sensor," *Journal of Microelectromechanical systems*, vol. 9, pp. 303–308, September 2000.

[198] D. Briand, H. Sundgren, B. van der Schoot, I. Lundstrm, and N. de Rooij, "Thermally isolated MOSFET for gas sensing application," *IEEE Electron Device Letters*, vol. 22, pp. 11–13, January 2001.

[199] W. S. Ruska, *Microelectronic processing. An introduction to the Manufacture of integrated circuits*. McGraw-Hill ed., 1988.

[200] F. Iker, "Low temperature wafer bonding techniques for MEMS and IC," Master's thesis, Universit Catholique de Louvain, Louvain-La-Neuve (Belgium), June 2002.

[201] A. Bosseboeuf and S. Petitgrand, "Characterization of the static and dynamic behaviour of M(O)EMS by optical techniques: status and trends," *Journal of Micromechanics and Microengineering*, vol. 13, pp. S23–S33, 2003.

[202] K. J. Gasvik, K. Creath, K. M. Crennell, N. A. Halliwell, C. J. Pickering, P. Hariharan, M. Kujawinska, and T. Yatagai, *Interferogram analysis*. Bristol and Philadelphia: David W. Robinson and Graeme T. Reid, Institute of Physics Publishing ed., 1993.

[203] V. Quintard, *Rflectomtrie et interfromtrie trs haute rsolution. Application la caractrisation de composants lectroniques*. PhD thesis, Universit de Bordeaux, Bordeaux, France, 1995. nbr. 1519.

[204] E. Schaub, *Etude par thermorflectivit du comportement de diodes laser de puissance pour tlcommunication*. PhD thesis, Universit de Bordeaux, Bordeaux, France, 1999. nbr. 2149.

Publications originated from this work

Some pictures of this book have been reprinted from previous first author publications with permission of their editors, IEEE, Elsevier, SPIE, AIAA and IOP.

1. J. Laconte. Etude et ralisation de structures pour capteurs intgrs et circuits micro-ondes par micro-usinage du Silicium au TMAH. Master's thesis, Microelectronics Laboratory of the Universit catholique de Louvain, Louvain-La-Neuve, September 1999.

2. J.-P. Raskin, J. Laconte, A. Akheyar, S. Adriaensen, A. Nve, I. Martinez, M. Dehan, B. Parvais, D. Vanhoenacker, L. Demeus, P. Delatte, V. Dessard, and D. Flandre. Fully-depleted SOI CMOS Technology for Heterogeneous Micropower, High-Temperature or RF Microsystems. *Belgian Journal of Electronics and communications*, (2):53–68, July 2001.

3. D. Flandre, S. Adriaensen, A. Akheyar, A. Crahay, L. Demes, P. Delatte, V. Dessard, B. Iniguez, A. Nve, B. Katschmarskyj, P. Loumaye, J. Laconte, I. Martinez, G. Picun, E. Rauly, C. Renaux, D. Spte, M. Zitout, M. Dehan, B. Parvais, P. Simon, D. Vanhoenacker, and J.-P. Raskin. Fully-depleted SOI CMOS Technology for Heterogeneous Micropower, High-Temperature or RF Microsystems. *Solid-State Electronics (Elsevier Science, Pergamon)*, 45(4):541–549, April 2001.

4. J. Laconte, C. Dupont, A Akheyar, J.-P. Raskin, and D. Flandre. Fully CMOS compatible low-power microheater. In SPIE, editor, *Design, Test, Integration and Packaging of MEMS/MOEMS, DTIP 2002*, volume 4755, pages 634–644, Cannes Mandelieu, Cte d'Azur, France, May 5-8 2002.

5. D. Flandre, S. Adriaensen, A. Afzalian, J. Laconte, D. levacq, C. Renaux, L. Vancaillie, and J.-P. Raskin. Intelligent SOI CMOS integrated circuits and sensors for

heterogeneous environments and applications. In *IEEE Sensors 2002 Conference*, pages 1407–1412, Orlando, Florida, USA, June 12-14 2002.

6. J. Laconte, C. Dupont, D. Flandre, and J.-P. Raskin. SOI CMOS compatible low-power microheater optimization and fabrication for smart gas sensor implementations. In *IEEE Sensors 2002 Conference*, pages 1395–1400, Orlando, Florida, USA, June 12-14 2002.

7. S. Jorez, J. Laconte, J.-P. Raskin, A. Cornet, S. Grauby, S. Dilhaire, and W. Claeys. Optical characterization method for MEMS. In *Photonics and Mechanics Conference, Photomec'03*, Louvain-La-Neuve, Belgium, February 21 2003.

8. J. Laconte, V. Wilmart, J.-P. Raskin, and D. Flandre. Capacitive humidity sensor using a Polyimide sensing film. In IEEE, editor, *Design, Test, Integration and Packaging of MEMS/MOEMS, DTIP 2003*, pages 223–228, Cannes Mandelieu, Cte d'Azur, France, May 5-7 2003.

9. R. Charavel, J. Laconte, and J.-P. Raskin. Advantages of p++ polysilicon etch stop layer versus p++ silicon. In SPIE, editor, *Smart Sensor, Actuators, and MEMS Conference*, volume 5116, pages 699–709, Gran Canaria, Spain, May 19-21 2003.

10. J. Laconte, V. Wilmart, and J.-P. Raskin. High-sensitivity capacitive humidity sensor using three-layer patterned Polyimide sensing film. In *IEEE Sensors 2003 Conference*, pages 372–377, Toronto, Canada, October 21-24 2003.

11. P. Ivanov, J. Laconte, J.-P. Raskin, M. Stankova, E. Sotter, E. Llobet, X. Vilanova, D. Flandre, and X. Correig. SOI CMOS compatible low-power gas sensor using sputtered and drop-coated metal-oxide active layers. In IEEE, editor, *Design, Test, Integration and Packaging of MEMS/MOEMS, DTIP 2004*, pages 137–142, Montreux, Switzerland, May, 12-14 2004.

12. J. Laconte, F. Iker, S. Jorez, N. Andr, T. Pardoen, J. Proost, D. Flandre, and J.-P. Raskin. Thin films stress extraction using micromachined structures and wafer curvature measurements. In *Conference on advanced Microelectronics Materials, Materials for Advanced Metallization, MAM'04*, Brussels, Belgium, March, 8-10 2004.

13. J. Laconte, F. Iker, S. Jorez, N. Andr, T. Pardoen, J. Proost, D. Flandre, and J.-P. Raskin. Thin films stress extraction using micromachined structures and wafer curvature measurements. *Microelectronic Engineering Journal*, 76:219–226, 2004.

14. J. Laconte, C. Dupont, D. Flandre, and J.-P. Raskin. SOI CMOS compatible low-power microheater optimization for the fabrication of smart gas sensors. *IEEE Sensors Journal*, 4(5):670–680, October 2004.

15. J. Laconte, B. Rue, D. Flandre, and J.-P. Raskin. Fully CMOS-SOI compatible low-power directional flow sensor. In *IEEE Sensors 2004 Conference*, pages 864–867, Vienna, Austria, October 24-27 2004.

16. P. Ivanov, J. Laconte, J.-P. Raskin, M. Stankova, E. Sotter, E. Llobet, X. Vilanova, D. Flandre, and X. Correig. SOI-CMOS compatible low-power gas sensor using sputtered and drop-coated metal-oxide active layers. *Journal of Microsystem Technologies (Springer)*, Accepted for publication, To be published.

17. D. Flandre, J. Laconte, D. Levacq, A. Afzalian, B. Rue, C. Renaux, F. Iker, B. Olbrecht, N. Andr, and J.-P. Raskin. SOI technology for single-chip harsh environment microsystems. In AIAA, editor, *Conference on Micro-Nano-Technologies for Aerospace Applications, CANEUS*, pages 157–169, Monterey, California, USA, October 30 - November 5 2004.

18. S. Jorez, J. Laconte, A. Cornet, and J.-P. Raskin. Low cost instrumentation for MEMS thermal characterization. *Journal of Measurement Science and Technology, IOP, Institute of Physics*, Accepted for publication, To be published.

19. J. Laconte. *Micromachined Thin-Film Sensors for SOI-CMOS Co-integration*. PhD thesis, Universit catholique de Louvain, Louvain-La-Neuve, Belgium, November 2004.

Index

acetone vapor, 208
adhesion layer, 38, 196, 198–200
ageing, 150, 181, 183, 207
alkali ions, 20, 22, 219
ammonia vapor, 208
ammonium persulfate, 31
anemometer, 187
anisotropic etchants
    EDP, 20, 22
    KOH, 20, 22
    TMAH, 17, 22
anisotropic etching, 19
ANSYS simulations, 115

bending moment, 58
biaxial modulus, 50
body effect, 217
boron, 55
    etch stop, 22, 39
buckling, 70
    critical buckling, 71
    post-buckling, 73
bulk micromachining, 5, 17, 19

calibration
    flow sensor, 168
    microheater, 129
calorimetric flow sensor, 163, 176
cantilever, 42, 86
    monolayer, 86
    multilayer, 94
carbon monoxide, 207
Chromium, 38, 196, 197
CMOS compatibility, 3, 7, 8, 11, 18, 22, 31, 36, 108, 188, 196, 198, 199, 203, 213, 251
co-integration, 3, 8, 123, 214, 235, 241, 251
concave corner, 41
conduction, 113, 116, 117
convection, 114, 117
convex corner, 41, 71, 77, 84, 89
critical point drying, 93
crystallography, 255

Deep Reactive Ion Etching (DRIE), 19
densification, 64
dicing, 126
direct tunneling, 217

e-beam evaporation, 198, 219, 227
EDP, 20, 22
elastic constants, 51
elastic modulus, 50
etch stop, 38
    boron, 22, 39
    dielectric, 39
    electrochemical, 40
ethanol vapor, 208

flow sensor, 163, 235

anemometer, 187
    calorimetric, 163, 176
    time-of-flight, 187
flow velocity, 163, 175–177, 179, 181, 183–185, 188
forming gas, 198, 219
Fowler-Nordheim tunneling, 217

gas sensitive layer deposition, 200
    drop coating, 200, 201
    screen-printing, 200, 202
    sputtering, 200, 201
gas sensor, 107, 110, 193, 242
gasFET, 242
gated diode, 232
Gold, 22, 30, 38, 197

heat transfer, 113
    conduction, 113
    convection, 114
    radiation, 115
HNA, 19
hybrid technology, 3, 8

interdigitated electrodes, 110, 141, 195, 205
interferometry, 71, 78, 90, 158, 259
isotropic etchant
    HNA, 19
isotropic etching, 19

kink effect, 216
KOH, 20, 22

lattice
    cubic, 256
    diamond, 256
lift-off, 196–199, 201, 204, 220

membrane, 6, 111, 251

design and simulation, 116, 119
    dimensions, 133
    patterning, 43
    reliability, 158
    residual stress, 47, 65, 67, 96
    thermal uniformity, 138
    thickness uniformity, 38
    transistors on membrane, 235
    yield, 126
mesa structure, 42
methane, 207
microheater, 112, 115, 127, 136, 163, 205
    accesses, 127, 136–140, 142
    hot spot, 112, 115, 130, 139
    loop-shape, 112, 116, 127
    meander, 115, 127
microhotplate, 107, 193
micromachined structure
    cantilever, 42, 86
    clamped-clamped beam, 71, 99
    microgauge, 82, 98, 99
    ring-and-beam, 71
microsystem, 3, 10
Miller indices, 255
misalignment, 41, 43, 44, 235
moment of inertia, 58
MOS capacitor, 230
MOS transistor, 222
    on membrane, 235

nitrogen dioxide, 207
nucleation, 67

oxide charge and interface trap
    fixed oxide charge, 218
    interface charge trap, 219, 227
    mobile ionic charge, 219

INDEX

oxide trapped charge, 219, 227
oxynitride, 65

packaging, 9, 126, 175
Platinum, 38, 112, 197, 198
Poisson ratio, 51, 52
Polyimide, 7, 30
polysilicon
    at high temperature, 142
    grain, 148
    phosphorous doped, 55, 112, 120
    residual stress, 99
    Seebeck coefficient, 166, 171
    sheet resistance, 112, 120, 129
    TCR, 112, 113, 172
    thermal conductivity, 117
    thermal stability, 113, 142, 150
    thermopiles, 166
    thermoreflectivity, 138
post-processing, 17, 36, 121, 126, 194, 204, 220, 222, 224, 225, 229–233, 235, 251
power consumption, 111, 117, 133–137, 145, 146, 150, 151, 153, 156–158, 169, 185, 205

radiation, 115
reflectometry, 138, 154, 171, 263
reliability, 158
response time, 173, 179, 185

sacrificial layer, 5, 99
screen-printing, 200, 202
scriber, 126
Seebeck
    coefficient, 165, 171
    effect, 165

self-heating, 217, 240
sensor, 3
    flow, 163
    gas, 107, 110, 193, 242
    humidity, 7
    IR, 111
    pressure, 6, 241
    smart, 3, 110
silicates, 30
silicic acid, 31
silicon crystallography, 255
Silicon-on-Insulator (SOI), 4, 39, 110, 122, 215, 216
    buried oxide stress, 65
    fully-depleted, 215
    partially-depleted, 215
    SIMOX, 215
    Smart-Cut, 215
    UNIBOND, 215
silicon-rich nitride, 65
SnO_2, 110, 119, 193, 201, 209
stiction, 93, 101
stoichiometric nitride, 51, 61
Stoney
    equation, 59, 63
    superposition principle, 61, 66
strain, 49, 52
    compressive, 71, 82
    gradient, 95
    measurement techniques, 69
    relaxed, 52
    tensile, 73, 82
stress, 48
    aluminum, 98
    average, 52
    biaxial, 50

compressive, 47, 54
extrinsic, 53
gradient, 52, 66, 86
intrinsic, 55
measurement techniques, 56
nitride, 102
oxide, 102
plane, 49
polysilicon, 99
residual, 47, 49
shear, 49
tensile, 40, 47, 54
thermal-mismatch, 53
uniaxial, 50
stress and strain measurement
micromachined structures, 69
substrate curvature, 56, 68, 69, 88
supercritical region, 93
surface micromachining, 5, 19

temperature coefficient of resistance (TCR), 112, 130, 131, 136, 143, 149, 151, 172
thermal conductivity, 113, 114, 117, 118, 135
thermal expansion coefficient, 54
thermal inertia, 173
thermal uniformity, 107, 110, 112, 116, 119, 138
thermopiles, 164
thermoreflectometry, 138
time-of-flight, 187
Titanium, 38, 197, 199
TMAH, 17, 22, 71, 76, 84, 89, 98, 121, 123, 126, 196, 198, 199, 204, 220, 222, 223, 230, 233, 234, 238, 240

aluminum selectivity, 29
dielectrics selectivity, 28
etch rate, 25
etch stop, 22, 38
metals selectivity, 37
silicon etch rate, 22, 23, 34
silicon roughness, 22, 27, 34
undercutting, 41, 71, 77
transconductance over drain current ratio, 218
trimming, 148
Tungsten, 38, 197, 198

WO_3, 110, 193, 201, 209

X-ray, 198, 219, 227

yield, 9, 126
Young's modulus, 50–52, 94